Project AIR FORCE

T0166311

STRATEGIC APPRAISAL 1996

Edited by

Zalmay Khalilzad

Prepared for the United States Air Force

RAND

Strategic Appraisal 1996 deals with the major strategic issues confronting the United States in the post–Cold War era, with an emphasis on the future role of aerospace power. It discusses both the question of U.S. grand strategy for the new era and the trends in various regions of the world and the challenges they could pose to the military forces of the United States—especially the Air Force.

The context for understanding our strategic situation is provided by a discussion of alternative U.S. grand strategies for the post–Cold War era. Despite the amount of time that has passed since the fall of the Berlin Wall and the dissolution of the Soviet Union, the nation has not reached a consensus on a replacement for the containment strategy that played such a large role in informing our international behavior over decades. In the absence of such a grand strategy, we will continue to have difficulty in assessing the importance of international events for our long-term interests and the wisdom of using U.S. military power to affect them.

The discussion of the geopolitical trends in the various regions of the world is undertaken with an eye toward the potential challenges to the United States and its aerospace power that they might pose. With respect to each of the regions considered, the political situation is related to U.S. interests and to the possibility of U.S. military involvement.

Drawing primarily on the expertise of RAND researchers in a wide variety of geopolitical issues, this book was produced in the Strategy and Doctrine Program of RAND's Project AIR FORCE, which is sponsored by the United States Air Force. Although the work it contains was originally undertaken for Air Force leaders and planners, it should be of interest to others interested in national security issues, students of international relations, and those interested in the future of aerospace power.

PROJECT AIR FORCE

Project AIR FORCE, a division of RAND, is the Air Force federally funded research and development center (FFRDC) for studies and analyses. It provides the Air Force with independent analyses of policy alternatives affecting the development, employment, combat readiness, and support of current and future aerospace forces. Research is being performed in three programs: Strategy and Doctrine, Force Modernization and Employment, and Resource Management and System Acquisition.

CONTENTS

FIGURES

TABLES

INTRODUCTION
Zalmay Khalilzad

The Western victory in the Cold War has left the United States as the world's preeminent power. At the moment, it faces no global rival and no significant hostile alliance that might threaten its security or vital interests. Despite a decline in its relative economic power, the United States still has the world's largest economy. Moreover, the United States possesses military predominance, and American political and economic ideas have broad global appeal. Almost all of the economically capable nations are our allies. In modern times no single nation has held a position as preeminent as that of the United States today.

At the same time, with the end of the Cold War, there has been an increase in disorder as a result of the rise in ethnic nationalism and the fragmentation of several states. Among the Cold War allies, there is intense economic competition, and managing alliance relations has become more difficult in the absence of a powerful common enemy. Several powers are opposed to the new international configuration. Some of these are rogue states, such as North Korea, Iran, and Iraq. But some major powers, such as Russia and China, are also dissatisfied with the *status quo.*

NATURE OF THE PROBLEM

During the Cold War, the United States had a grand strategy—containment—which guided its actions. But this coherent U.S. strategy collapsed along with the Berlin Wall. Lacking a global enemy, the United States has been operating without a grand strategy for the new era. Despite efforts by parts of the Bush and Clinton administrations, no grand strategy has yet jelled. Although lacking a broadly accepted grand strategy, the United States has a national military strategy that focuses on regional threats—especially in Korea and the Persian Gulf—and the capability to fight two major regional contingencies (MRCs) nearly simultaneously. The defense budget remains under pressure, and technological changes under way pose major new challenges.

There are a number of problems with a strategy that focuses on two MRCs. While in the post–Cold War era regional threats are indeed important, the focus on Korea and the Gulf for planning purposes raises a number questions. First, it does not deal with Europe. Second, although lesser regional conflicts are dominating our use of force, the current military strategy does not provide guidance as to whether and how we should engage our forces in any particular crisis. Third, in the long run, the focus on the Gulf and Korea has the potential to miss systemic changes in the international security environment. Russia could revert to an aggressive international posture, or China might become expansionist as its economy and military might grow. In any case, a gap may be emerging between the military strategy and the capabilities available to carry it out. This gap could grow if budget constraints increase—which is likely, given the growing emphasis on balancing the budget within the next seven years.

The focus on the Korean and Persian Gulf MRCs not only underestimates these other challenges but could have important consequences in the long run for political support for an adequate defense budget. In either or both of these regions, the political circumstances that give rise to the MRC threat could change quickly: Korean unification could occur suddenly, and the regimes of both Iran and Iraq are under great stress. While the disappearance of these trouble spots would, of course, affect U.S. military requirements, it would not abolish them, as the current strategy might seem to suggest. Therefore, a review of the national military is likely in the near future—most likely after the next presidential election.

PURPOSE OF THE BOOK

The work contained in this book was undertaken to help the Air Force understand the strategic background against which these issues must be addressed. Two aspects of this strategic background are considered. First is the question of U.S. global strategy. Given its preeminent position of power, the single most important determinant of the future international environment is the set of choices that the United States itself makes about its role in the world, its objectives, and the extent to which it wishes to be proactive in shaping the international system. Obviously, the choices the United States makes at this level will also affect its defense budget and the type and size of military establishment that it will need to maintain in the future.

However, these choices will not be made in a vacuum. The post–Cold War world will present a series of crises, challenges, and opportunities that will at times limit U.S. options and force U.S. decisions. While U.S. strategy will affect whether and how vigorously we choose to respond, the pressures for involvement will exist and, from time to time, become large and perhaps overwhelming. A review of the major regions of the world illustrates the kind of challenges that will be presented and to which U.S. military force, in general, and the U.S. Air Force, in particular, may be called upon to respond.

If a consensus were to emerge with respect to a grand strategy for the post–Cold War era, it would, of course, have a major effect on defense planning. Such a decision would be a major determinant of defense budget levels, as well as force structure and the balance between current readiness, on the one hand, and procurement and research and development, on the other. However, despite its importance, such a political consensus on global strategy may not emerge any time soon. Thus, the U.S. armed forces must plan with an eye toward the kinds of challenges that will emerge and to which, regardless of grand strategy, they may be called upon to respond.

STRUCTURE OF THE BOOK

In accordance with the above framework, the book is organized as follows:

Part One discusses the need for a national grand strategy and describes three possible strategies and their implications for global stability and national military strategy. These global goals provide filters for assessing the trends in the various regions of the world in terms of their importance to the United States:

- **Neoisolationism.** This option would involve abandoning U.S. global preeminence and turning inward to face domestic problems. Although this approach could produce significant defense savings and other benefits in the short run, it would likely increase the danger of major conflicts and, hence, would require much greater U.S. defense efforts over the long term. Thus, it could eventually undermine U.S. prosperity.

- **Multipolarity.** This option would rely on the balance of power among several nations to preclude the emergence of a superpower competitor to the United States. As in the 19th century, the great powers would alternately compete and cooperate to avoid hegemony and global war. There could be advantages for the United States in a such a strategy, including a reduction in the defense burdens, albeit smaller than that associated with a neoisolationist strategy. The risks, however, could be severe. These include the possibility that the United States would face increased competition from other major powers; that a decline in U.S. influence might have negative economic consequences, including a weakening of the General Agreement on Tariffs and Trade and the International Monetary Fund; that the members of such a system would find it harder to behave according to its rules than was the case in the ultimately failed 17–19th century system; and that such an international system would lead to new arms races and even world wars.

- **Global leadership.** In this option, the United States would seek to maintain a position of global leadership, thereby precluding both the rise of another global rival and multipolarity. To succeed in the long term in realizing this overall objective, the United States would have to adhere to seven principles as guidelines for its policies:

  — Maintain and selectively extend the network of alliances and cooperation among the economically most capable democratic nations.

— Preclude hostile hegemony over critical regions.

— Hedge against Russian reimperialization and Chinese expansionism while promoting cooperation with both.

— Preserve U.S. military preeminence by maintaining the right force size and mix.

— Maintain U.S. economic strength and an open international economic system, and reduce the social crisis in our country.

— Be judicious in the use of force, avoid overextension, and achieve effective burden-sharing with allies.

— Obtain and maintain domestic support for global leadership and a strategy able to support it.

The key question is whether the American political leadership and people will support such a strategy. The near-term costs of this strategy will be higher in terms of resources for defense than if the country adopts one of the other two alternatives. However, it is possible that the public might support it if it were presented by the president and supported by the senior members of both parties and if the costs and benefits of such a strategy and the alternatives were debated and understood. The costs of alternative approaches to U.S. global leadership can ultimately be higher. In addition, there are economic benefits for the United States from playing a global leadership role. These benefits have not been focused on, either analytically or in public debate. Global leadership and building a more democratic and peaceful world should also appeal to American idealism, a defining American feature. For sustaining domestic political support, this appeal might well be as important as more selfish and material American interests. In fact, having such a goal can be a spur to the kinds of social and educational reforms that are necessary, a point missed by those who argue that we should tend to our own problems rather than be concerned about the rest of the world.

Part Two provides an assessment of current geopolitical trends, including a description of the demands that might be placed on the U.S. military and, in particular, on the Air Force. As noted, these demands can serve as guideposts to future military requirements, even in the absence of a new grand strategy. These trends, and their implications for the possible use of U.S. aerospace power, are discussed in a series of chapters dealing with the various regions of the world. Several overall themes emerge from these chapters:

- In many parts of the world, an underlying trend toward democracy and free markets remains evident, despite buffeting and crosswinds. The U.S. armed forces have a role to play in furthering this trend, through military-to-military contacts, joint exercises, and other operations. This is especially true in those regions where the military itself has often been an opponent and obstacle to democratization; both as a model and as a source of military expertise, professionalism, and prestige, the U.S. armed forces can have an important positive influence on those military establishments.

- In some areas, however, notably the Persian Gulf and the Korean peninsula, U.S. military power remains an essential factor in deterring aggression against allies and friends. Aerospace power, which can both be exerted from outside the region and be rapidly deployed to it, will be an increasingly important part of that deterrent capability.

- Over the longer term, several important strategic questions may arise that will have major military dimensions:

  — In East Asia, if China successfully manages the post-Deng transition and continues to grow economically at anything like its current rate, it will be in a position to become a major military force. Depending on the U.S. view of China's role in the international system, the United States could confront the problem of maintaining access to the world's economically most dynamic region in the "shadow" of Chinese economic and military power.

  — Related to this will be the need to define a security relationship with a reunified Korea. As the proverbial "minnow among the whales," Korea may wish to maintain a strong military relationship with the United States to counterbalance the power of both China and Japan. However, the continued stationing of U.S. troops may be resisted on nationalist grounds.

— In Europe, the United States and its NATO allies have taken on the responsibility for enforcing peace in Bosnia. The outcome of this commitment can have a lasting effect on U.S.–European security ties. The United States might take on additional military responsibilities as NATO expands eastward—both to stabilize East-Central Europe and as a hedge against Russia going bad. How NATO's eastward expansion is related to European Union and Western European Union expansion and future ties with Russia will be a central challenge of the coming era. Whatever the modalities of expansion, it will place new demands on the U.S. armed forces, especially the Air Force.

- The proliferation of weapons of mass destruction (WMDs) will continue to pose a threat to U.S. national security. The full range of foreign policy tools will be required to deal with it. Military requirements may include the ability to attack WMD facilities or stockpiles, defense against ballistic missiles and other WMD delivery systems, and the capability to intervene to prevent the use of WMDs in a regional conflict or to deal with the aftermath.

- The world's recent bout of post–Cold War disorder is likely to continue and possibly to spread. U.S. military power may be called upon for peacekeeping or peacemaking duties, or to undertake humanitarian relief efforts that surpass, in terms of scale or difficulty, the capabilities of civilian governmental and quasi-governmental agencies. The use of aerospace power in operations of this sort raises different issues than does its use in more conventional types of conflict. The United States lacks clear guidelines for involvement in these lesser regional conflicts—whether conducted unilaterally, as part of an *ad hoc* coalition, or as part of a United Nations operation.

In sum, the United States, including the Air Force, is in a position of unusual preeminence. The key challenge for the United States is to preserve that position by maintaining the technological superiority of U.S. weapons and the quality of the people who serve in the armed forces, and avoiding the mistakes of the past when the armed forces were cut too rapidly; by using its armed forces selectively in defense of vital American interests, defined in terms of preserving its current global status; and by promoting long-term regional security. Meeting this challenge will be a tall order for the country's political and military leadership.

Chapter One

U.S. GRAND STRATEGIES: IMPLICATIONS FOR THE UNITED STATES AND THE WORLD

Zalmay Khalilzad

Where is the United States going?

During the Cold War, U.S. foreign and security policy policies were guided by the doctrine of containment. Five years after the end of the Cold War, however, no new paradigm or grand design has emerged. Does the country need a new vision and a new grand strategy? What are the opportunity costs of not having a new grand strategy? What alternatives does the country face? And what are the implications of different options for U.S. foreign and security policies and U.S. military forces—including the Air Force?

NEW INTERNATIONAL STRUCTURE

The end of the Cold War produced the second extraordinary change in the global balance of power in this century. In its first 50 years, there were two world wars and two major revolutions, in Russia and in China. Five empires collapsed—the Ottoman, the Austro-Hungarian, the German, the Italian, and the Japanese. Two other global imperial systems—the British and the French—declined dramatically. As a result, the character of the international system underwent fundamental change. For several centuries, the international order had been characterized by multipolarity and a balance of power. No single nation was allowed to gain such predominance that a coalition of other states could not confront it with greater might. The system succeeded in preventing the emergence of a single dominant power, but it ultimately failed and led to World War I, which was followed by a chaotic period, the rise of fascism, and World War II. This war was followed by the emergence of a global bipolar system.

The transformation to bipolarity occurred for two reasons. There was a dramatic decline in the relative power of several key members of the old (pre-World War I) balance-of-power system. Germany was defeated in World War II. Britain and France experienced a significant decline. These developments coincided with another important change: the concentration of relative power in the United States and the Soviet Union and their active engagement in global affairs. These changes, which were the result of a complex set of factors, produced a new international system. A special feature of the new system was the fact that the Soviet Union and the United States represented two different value systems and ways of life—issues different from those that had driven the conflicts in the multipolar balance-of-power era. Moscow was animated by a revolutionary ideology and a sense of historic mission. After a period of uncertainty, the United States decided to undertake a determined effort to contain the spread of Soviet power. This struggle, the Cold War, took place in the context of the development and refinement of weapons of mass destruction (WMDs), with the ever-present danger of nuclear annihilation.

The Cold War dominated U.S. foreign policy, national security strategy and major defense decisions—weapon system acquisition, force sizing, overseas presence, and alliances. Cold War bipolarity required the United States to be prepared to contain the spread of Soviet power on a global basis. This principle affected U.S. dealings with various regions.

The Cold War ended with the sudden collapse of both the Soviet empire and the Soviet state. As a result, the relative balance of political and military power shifted in favor of the United States. Through the more than four decades of the Cold War, the United States accumulated enormous political status and vast military capabilities. At the moment, the United States faces no global rival and no significant hostile alliances. Most economically capable nations, including those with both high per capita and high total Gross National Product (GNP)—such as Germany and Japan—are U.S. allies. Indeed, the United States and its democratic allies may be said to have constituted a "zone of peace" in which war among them is "unthinkable." However, as far as economic indicators of relative power are concerned, although the United States remains the world's largest economy, there has been a decline in its relative economic power. The country also faces significant domestic social problems, resulting in an increased domestic focus on the part of the political elite.

Besides America's sole-superpower status, there are seven other important features of the current international environment. First, dramatic economic growth is under way in Asia, in countries such as China, India, Indonesia, and Thailand. If current trends hold—a questionable proposition—it will produce a shift in relative economic power with important potential geopolitical and military implications.

Second, significant parts of the rest of the world, such as Latin America, East-Central Europe, and the Middle East are experimenting in market economies and democratic government. Some are likely to succeed.

Third, much of the rest of the world is an undemocratic zone of conflict where there is the danger of MRCs, attempts at regional hegemony, and proliferation of weapons of mass destruction and the means to deliver them over increasingly longer distances.

Fourth, there is also an increasing risk of chaos and fragmentation within states due to political decay and to ethnic, sectarian, and ideological factors, which can produce mass starvation and attempts at genocide with humanitarian and, at times, geopolitical implications. This means that the United States and other members of the zone of peace and prosperity are likely to be confronted by a significant and perhaps growing number of small wars.

Fifth, important and accelerating technological changes are under way with dramatic potential effects on the global economy and military power. Those who make the right choices will have comparative advantage over others. On the military front, the choices not only involve new weapons—in areas such as information, stealth, and precision guidance—but also new concepts of operation and organizational changes.

Sixth, there are intensified international economic competition and trade frictions between the United States and its Cold War allies, which may complicate future relations.

Seventh, a number of states, such as Iran, North Korea, Cuba, Iraq, China, and Russia, are unhappy with the current global system and over the longer term—the next 20 years—there is a real possibility of efforts by China or Russia or a coalition of states to balance the power of the United States and its allies.

The interaction of these factors is likely to determine the geopolitical shape of the world in the 21st century. They will be the sources of future conflicts and alliances. Which of the trends become dominant and which ones come together will determine future conflicts. Different combinations of these trends will produce different types of conflicts not only in terms of their causes but also in terms of the ways in which they are fought and in terms of coalition states cooperating or competing. The spectrum of possibilities is quite wide. For example, information technologies—which are changing and will continue to change dramatically well into the next century—alone are likely to affect the spectrum of conflict from major wars to terrorist actions. Different combinations of sources and means used will place very different demands on U.S. armed forces. The current U.S. alliances may further change, and new alliances may come about.

Although understanding the range of plausible future conflicts is important, equally important is estimating the time frame for the emergence of various conflicts. For example, in the immediate future, it may be that lesser regional conflicts (LRCs)—ethnic conflicts, civil wars, peacekeeping, and peacemaking—will be the dominant type of conflict. Over time, and depending on the changes in the world and perceptions about the United States, other forms of conflicts—major regional war or even global war—could become more plausible.

Demand alone is unlikely to determine whether the United States will get involved in future conflicts. As history indicates, the United States will not get involved in all conflicts. Given the budgetary trends, the United States is unlikely to have the capability to deal with the whole spectrum of threats to international peace and stability by itself. Whether the United States would get involved by using its armed forces will also depend on its perception of what is at stake for it. That, in turn, will depend on U.S. grand strategy and U.S. national security strategy. These strategies can shape the role that the United States plays in the world and how various crises and conflicts are evaluated.

ALTERNATIVE GRAND STRATEGIES

Given the changes in the global scene, the United States faces fundamentally different challenges and choices than it did during the Cold War. Arguably, what the United States chooses will set not only its own direction, but also that of the rest of the world for the next century. However, five years after the end of the Soviet Union, there is no consensus in the United States on a new vision and a grand strategy for the new era.

A strategic vision and a grand strategy are important because they will provide the United States a strategic direction that will guide long-range planning in the Department of Defense and the services. Without it, the country and its armed forces are likely to be placed in a reactive mode, losing the initiative to others. Given America's power position in the world, the United States is in a position to shape the future of the world. But it cannot succeed in shaping the post–Cold War world unless it knows what shape it wants the world to take and has the strategy and the will to make it happen.

Without a vision and a grand strategy, the country will also lack a standard for judging conflicts and crises that do not threaten the U.S. homeland. Specific policy decisions cannot be evaluated adequately without first constructing a framework for guiding policy and setting priorities. Absent such a framework, it will be more difficult to decide what is important and what is not, to determine which threats are more serious than others. A vision and a strategy are particularly important now for another reason. Given the budgetary trends, there is a need for prioritization as the armed forces downsize. Depending on the national strategy, the emphasis between, for example, readiness and modernization, could change.

The shift in the tectonics of power confronts Washington with several options. It could abandon global leadership and turn inward. It could seek to give up leadership gradually and facilitate the trend toward diffusion of power by reducing the U.S. global role and encouraging the emergence of a 17th-to-19th century style balance-of-power structure with spheres of influence. Alternatively, the central strategic objective for the United States could be to consolidate U.S. global leadership and preclude the rise of a global rival.[1]

Neo-Isolationism

Abandoning predominance and turning inward could result in a significant reduction in defense expenses in the short run—although how much money the United States would, in fact, save over either the short or the long run should it adopt such a strategy has not been seriously studied. To assess how much money might be saved, the following questions would have to be addressed: Would U.S. defense include the defense of North America or the Americas generally? How far would the defensive perimeter extend into the Atlantic and Pacific Oceans?

Abandoning global leadership would also decrease the likelihood of placing American soldiers in harm's way around the world in places like Iraq, Haiti, Korea, Bosnia, and Somalia. The U.S. armed forces would not become involved in LRCs and MRCs in other parts of the world. Rather than expanding NATO, the United States would eliminate its military presence in Europe. Similarly, the United States would end its military presence in Asia and the Middle East. It would rely on space systems to gain situation awareness around the world. The United States would emphasize such capabilities as national missile defense, coastal defense, antisubmarine warfare and space warfare. It would also have to develop a spectrum of capabilities and strategies for protecting itself against information war. Programs to reduce the vulnerability of U.S. society and military forces to missile attacks and information warfare would become a high priority. Since the United States does not face a major ground threat to its territory, the army would shrink significantly in size.

---

[1]Several RAND analysts have debated and discussed alternative grand strategies for the United States. See Davis (1994a, pp. 135–164) and Levin (1994). Also see "Strategy and the Internationalists: Three Views" (1994). The broader community's debate has included Kennedy (1993), Huntington (1992), Krauthammer (1991), and "The Initial Draft of the Defense Department's Planning Guidance" (1992, p. 1).

The reduction in defense expenditure could help deal with the budget deficit and could improve U.S. economic competitiveness, especially since, at the same time, U.S. economic competitors might increase their defense expenditure. Ignoring foreign issues might enable the United States to pay more attention to domestic problems.

Furthermore, in some cases, allies to whose defense the United States has been committed need it less (e.g., the Soviet threat to Western Europe has disappeared, and the threats to Europe now are much smaller by comparison) and should be able to manage on their own (e.g., South Korea has more than twice the population and many times the GNP of North Korea). The U.S. defense commitment to the defense of its allies may enable them to spend less on defense.

Realistically and over the longer term, however, a neoisolationist approach might well *increase* the danger of major conflicts, require greater U.S. defense effort down the line, threaten world peace, and eventually undermine U.S. prosperity. By withdrawing from Europe and Asia, the United States would deliberately risk weakening the institutions and solidarity of the world's community of democratic powers, establishing favorable conditions for the spread of disorder—a return to conditions similar to those of the first half of the 20th century.

In the 1920s and 1930s, American isolationism had disastrous consequences for world peace. Then, the United States was but one of several major powers. Now that the United States is the predominant power, the shock of a U.S. withdrawal from the world could be even greater. In the extreme, U.S. withdrawal could result in the renationalization of Japan's and Germany's security policy in the long run. With U.S. withdrawal from the world, both Japan and Germany might have to look after their own security and build up their military capabilities. This can result in arms races, including the possible acquisition of nuclear weapons. China, Korea, and the nations of Southeast Asia already fear Japanese hegemony. Without U.S. protection, Japanese military capability would be likely to grow dramatically—to balance the growing Chinese forces and still-significant Russian forces. Given Japan's technological prowess, to say nothing of its stockpile of plutonium acquired in the development of its nuclear power industry, it could obviously become a nuclear weapon state relatively quickly, if it should so decide. Japan could also build long-range missiles and carrier task forces.

With the balance of power shifting among Japan, China, Russia, and potential new regional powers, such as Indonesia, Korea, and India, could come significant risks of preventive or preemptive war. Similarly, European competition for regional domination could also lead to major wars. If the United States stayed out of such a war—an unlikely prospect—Europe or Asia could become dominated by a hostile power. With domination of Europe or East Asia, such a power might well seek global hegemony, and the United States could face another global Cold War and the risk of another World War—one even more catastrophic than the last.

In the Persian Gulf, U.S. withdrawal is likely to lead to an intensified struggle for regional domination. Iran and Iraq have, in the past, both sought regional hegemony. Without American protection, the weak oil-rich states of the Gulf Cooperation Council (GCC) would likely not retain their independence. To preclude this development, the Saudis might seek to acquire, perhaps by purchase, their own nuclear weapons. If either Iraq or Iran controlled the region that dominates the world supply of oil, it could gain a significant capability to damage the U.S. and world economies. Whichever country were to gain hegemony would have vast economic resources at its disposal, which could be used to build military capability as well as gain leverage over the United States and other oil-importing nations. Hegemony over the Gulf by either Iran or Iraq would bring the rest of the Arab Middle East under its influence and domination because of the shift in the balance of power. Israeli security problems would increase, and the peace process would be fundamentally undermined, increasing the risks of war between the Arabs and the Israelis.

Already, rogue states, such as North Korea and Iran, are seeking nuclear weapons and long-range missiles. More states would acquire nuclear weapons if the United States became isolationist. Several states with potential nuclear capability, such as South Korea and Taiwan, have refrained from producing such weapons because of their security ties with the United States. Without such ties, these states and others might reconsider their nuclear posture. Similarly, those who are exercising restraint because they fear possible negative

U.S. actions are likely to be emboldened and shift to significant, perhaps overt, nuclear programs.

The extension of instability, conflict, and hostile hegemony in East Asia, Europe, and the Gulf would impact the American economy even in the unlikely event that the United States were able to avoid involvement in major wars and conflicts. Turmoil in the Gulf is likely to reduce the flow of oil and increase its price, thus reducing the American standard of living. Turmoil in Asia and Europe would force major economic readjustment in the United States, because it is likely to reduce the trading opportunities that have been so important to recent global prosperity, including the prosperity of the United States.

Return to Multipolarity and Balance of Power

A second option would rely on a balance of power to preclude the emergence of another predominant power. This approach has some positive features but is also dangerous. Based on current realities, the other potential great powers are Japan, China, Germany (or the European Union), and Russia, besides the United States. In the future, this list could change. New great powers—such as India, Indonesia, or Brazil—could emerge, or one of the existing ones—such as Russia—could decline or disintegrate and cease to be a great power.

The world may be heading toward a multipolarity. There are already several great powers in an economic sense, and the diffusion of wealth and technology will continue. Over time, the current economic powers might well decide to acquire political and military power commensurate with their economic strength. They are more likely to do so because in the post–Cold War world, the United States will not perceive threats in the same way as potential great powers, such as Germany and Japan—not to mention Russia, China, India, Brazil, Indonesia, etc.—and so would not be willing to run risks for them (Kissinger, 1994, p. 809).

The United States could seek to delay or to accelerate and facilitate such a development. If the United States adopts the objective of facilitating and even accelerating a return to multipolarity and a balance of power, Washington would seek to weaken NATO and seek to ultimately replace it (or in effect have it taken over) by the Western European Union (the military arm of the European Union) or to encourage individual European great powers to look after their own security. To push the world in this direction, the United States would end its presence in Europe as the Europeans built up their capability and a balance of power emerged on the continent. The United States could affect the pace of such a development by, for example, announcing that it intended to withdraw from Europe within a specific period of time—thus creating the framework for a European military buildup to balance Russia.

For such a balance-of-power system to work, either Germany would have to substantially increase its military power or the European Union would have to deepen and become a kind of superstate. The United States would continue to have a vital interest in preventing the domination of Europe—including Russia—by a single power. So, if the Germans decided to build up militarily, the United States would play its part by forming alliances with any European country or countries that sought to prevent German hegemony and by maintaining adequate forces in the United States and possibly in England. However, problems other than an attempt to establish hegemony over Europe, such as instability in the Balkans, East-Central Europe, or North Africa, would be the responsibility of the Europeans alone, and the United States would not get militarily involved in local conflicts in these regions.

Similarly, the United States would not get involved militarily on the territory of the former Soviet Union. In general, it would accept that these areas fell within the Russian sphere of influence. Western European powers, especially Germany, and Russia would have interests in East-Central Europe and would have to try to work out some rules for regulating their interactions.

In Asia, the United States would similarly become a balancer as China and Japan built up their capabilities. In the event of a serious imbalance between Japan and China, the United States could play a balancing role with forces based in the United States or possibly in some of the smaller states in the region. As in the case of Europe, the United States would seek to prevent the emergence of regional hegemony by shifting alliances; it would cooperate with other powers to protect common interests and be prepared to protect specific interests in the region, such as the lives and property of American citizens.

In the Persian Gulf, in this framework, the United States and other major powers would oppose the domination of the region by any one power, because such a power would acquire enormous leverage over states that depend on the region's oil. At the regional level, the United States and other major powers could rely on a balance between Iran and Iraq to prevent regional hegemony. Assuming the great powers were willing to pursue a joint policy toward the Persian Gulf, the fact that the United States is relatively less dependent on the Gulf than either Western Europe or Japan would give it a strong bargaining position when it came to allocating the burdens required by such a policy among the great powers. On the other hand, one or more great powers might be tempted to abandon the great power coalition and to support a potential hegemon in the Gulf in return for favorable access to the Gulf's resources and markets. Finally, the United States would have to be the dominant power affecting important security issues in the Americas.

Aside from the question of inevitability, a balance-of-power system would have some advantages for the United States. First, it could reduce defense expenditure (probably not, however, by as much as with a neo-isolationist strategy) and deploy American military force less often to world's "hotspots," since the United States would let other great powers take the lead in dealing with problems in their regions. The United States would focus its defense planning on deterring the emergence of a predominant power. The U.S. military would be primarily focused on confrontation among the great powers. Except in the Americas, the United States would not get involved in LRCs. The United States would not necessarily have to take the lead in places like Korea and the Gulf. For sizing purposes, it would aim to possess a force large enough to affect the balance between any two other great powers. Second, the United States would be freer to pursue its economic interests, even when they damaged its political relations with countries that had been, but were no longer, its allies, except in the particular case in which it required an alliance with one great power to ward off a bigger threat.

It is possible that, in a balance-of-power system, the United States would be in a relatively advantageous position compared to the other great powers. Given its relative distance from other power centers, the United States could try to mimic the former British role of an off-shore "balancer." As in the 19th century, the United States and other great powers would compete and cooperate to avoid hegemony and global wars. Each great power would protect its own specific interests and protect common interests cooperatively. If necessary, the United States would intervene militarily to prevent the emergence of a predominant power.

But there are also several serious problems with this approach. First, there is a real question whether the major powers would behave as they should under the logic of a balance of power framework. For example, would the Western European powers respond appropriately to a resurgent Russian threat, or would they behave as the European democracies did in the 1930s? The logic of a balance of power system might well require the United States to support a non-democratic state against a democratic one, or to work with one undesirable state against another one. For example, in the Gulf, to contain the power of an increasingly powerful Iran, the United States would have to strengthen Iraq. At times, the United States has been unable to behave in this fashion. For example, after Iraqi victory against Iran in 1988, balance-of-power logic would have demanded that the United States support strengthening Iran. However, because of ongoing animosity in U.S.-Iranian relations, the nature of Iran's regime, and moral concerns in U.S. foreign policy, Washington could not implement such a strategy. There are many other examples. Therefore, a grand strategy that requires such action is probably unrealistic.

Second, this system implies that the major democracies will no longer see themselves as allies. Instead, political and military struggles among them would become legitimate. Each would pursue its own economic interests much more vigorously and might well weaken economic institutions, such as the General Agreement on Tariffs and Trade (GATT) and the liberal world trade order. Such a development would increase the likelihood of major economic depressions and dislocations.

Third, the United States would be likely to face more competition from other major powers in areas of interest to the United States. For example, other powers might not be willing to grant it a "sphere of influence" in the Americas, but might seek, as Germany did in World War I, to reach anti-U.S. alliances with Latin American nations. As noted earlier, another great power might support a potential hegemon in the Persian Gulf.

Finally, the system might not succeed on its own terms. The success of the balance-of-power system requires that great powers maintain it without provoking war. Great powers must signal their depth of commitment on a given issue without taking irrevocable steps toward war. This balancing act proved impossible to perform even for the culturally similar and aristocratically governed states of 19th-century Europe. It is likely to prove infinitely more difficult when the system is global; the participants differ culturally; and the governments, because of the increasing influence of public opinion, are unable to be as flexible (or cynical) as the rules of the game in the balance-of-power system would require. Thus, miscalculations of the state of the balance could lead to wars that the United States might be unable to stay out of. The balance-of-power system failed in the past, producing World War I and other major conflicts. It might not work any better in the future, but war among major powers in the nuclear age is likely to be more devastating.

Global Leadership

Global leadership and deterring the rise of another hostile global rival or a return to multipolarity for the indefinite future is the third possible long-term guiding principle and vision. A world in which the United States exercises leadership is likely to have the following attributes. First, the global environment is likely to be more open and more receptive to American values—democracy, free markets, and rule of law. Second, such a world has a better chance of dealing cooperatively with the world's major problems, such as nuclear proliferation, threat of regional hegemony by renegade states, and low-level conflicts. Finally, such a world will avoid or at least delay another global Cold or Hot War and all its dangers.

To preclude the rise of another global rival or multipolarity, the United States would have to adhere to the following principles as guidelines for its policies:

- Maintain and strengthen the alliances among the economically rich democratic states of North America, Western Europe, and East Asia—the democratic "zone of peace"—and incrementally extend it.

- Prevent hostile hegemony over critical regions.

- Hedge against Russian reimperialization and Chinese expansionism while promoting cooperation with both.

- Preserve United States military preeminence.

- Maintain U.S. economic strength and an open international economic system.

- Be judicious in use of force, avoid overextension, and achieve effective burden-sharing among allies.

- Obtain and maintain domestic support for U.S. global leadership and these principles.

Maintaining the zone of peace requires, first and foremost, avoiding conditions that can lead to "renationalization" of security policies in key allied countries, such as Japan and Germany. The members of the zone of peace are in basic agreement and prefer not to compete with each other in *realpolitik* terms. But this general agreement still requires U.S. leadership. At present, there is greater nervousness in Japan than in Germany about future ties with Washington, but U.S. credibility remains strong in both countries. The credibility of U.S. alliances can be undermined if key allies, such as Germany and Japan, believe that the current arrangements do not deal adequately with threats to their security. It could also be undermined if, over an extended period, the United States is perceived as either lacking the will or the capability to lead in protecting their interests, or if increased economic and trade competition lead to trade wars.

In Europe, besides dealing with balancing Russian military might and hedging against possible Russian reimperialization, the near-term security threat to Germany comes from instability in East-Central Europe and, to a lesser degree, from the Balkans. For France and Italy, the threats come from conflicts in the Balkans and the danger of Islamic extremism, instability and conflict, and the spread of weapons of mass destruction and ballistic and cruise missiles to North Africa and the Middle East.

In East Asia, too, Japan favors an alliance with the United States to deal with uncertainty about Russia, growing Chinese military power, and ambition, and the threat of nuclear and missile proliferation on the Korean peninsula. Like Germany, Japan currently prefers to work with the United States to deal with these security problems. As long as U.S.-led allied and coalition actions protect their vital interests, these nations are less likely to look to unilateral means. This implies that the United States needs a military capability that is larger than might be required based on an isolationist or balance-of-power-based definition of U.S. interests.

Maintaining and adapting NATO and the U.S. alliance with Japan without the common Soviet enemy will be a major challenge. The Europeans and Japanese might not remain confident that the United States would play a leading role in protecting their interests. For the United States, playing such a role might become increasingly difficult if the costs of meeting future challenges to Asian and European security grow and if American public opinion perceives that the Europeans and Japanese are not doing their fair share to protect common interests. Increased trade problems with Europe or Japan or both could result in weakening and perhaps ultimately undermining security relations. Unless solutions are found to these problems, perhaps (and preferably) through the ultimate creation of a free-trade zone among the United States, the European Union (EU), and Japan, and unless new security and defense arrangements are made among the members of the zone of peace, there is the risk that the Japan-U.S. security alliance might break. If Japan or Germany (or the EU) goes its own way, prospects for the emergence of a multipolar system would increase.

Within the U.S. leadership framework, Europe's security interests can be best served if NATO remains the primary entity to deal with instability and conflict in the south and east and possible revanchism in Russia. To perform this role, NATO nations now appear willing to maintain a robust military capability as a hedge against Russia going bad and taking over such countries as Ukraine and the Baltic states, to prepare for the eventual membership of Central European nations in the alliance in coordination with EU expansion, and to develop more capability for power projection southward. Should this trend continue, the U.S. military services, including the Air Force, must prepare for new missions resulting from NATO expansion to the east and concerns about the south.

Asia has no NATO-like multilateral alliance. The core security relationships are the U.S.-Japanese and U.S.-South Korean ties. Maintaining security ties with Japan is important for both nations, even though trade relations between the two countries have a greater potential to create mutual antagonism than U.S.-German trade relations. The forces of U.S., Japanese, and South Korean security cooperation in Asia should deal with the threat from North Korea and hedge against uncertainties in Russia and China. The United States should maintain enough forces in the region for deterrence and, with reinforcements, the defense of critical American, Japanese, and Korean interests. At present, the main military threat is a possible North Korean attack against South Korea. This would, of course, change quickly if North Korea collapsed and the two countries became one: The end of North Korea would raise new questions about the sizing of U.S. forces for planning purposes generally and for Asia in particular.

If the United States decides to play a global leadership role, it would be vital for the United States to preclude hostile hegemony over critical regions. A region can be defined as critical if it contains sufficient economic, technical, and human resources so that a hostile power gaining control over it could pose a global challenge. Although this could change in the future, two regions presently meet the criteria: East Asia and Europe. The Persian Gulf is very important for a different reason—its oil resources are vital for the world economy. In the long term, the relative importance of various regions can change. A region that is critical to American interests now might become less important, while some other region might gain in importance. For example, Southeast Asia appears to be a region whose relative importance is likely to increase if the regional economies continue to grow as impressively as they have done in the past several years. The Gulf might decline in importance if cheaper alternative energy sources become available and the resources of the region therefore became less crucial to world prosperity.

At present, the risks of regional hegemony in Western Europe and East Asia are very small, and the United States is the predominant outside power in the Persian Gulf. In the near term—10 years—

Moscow is unlikely to pose a global challenge. However, even in its current weakened condition Russia could pose a major regional threat if it moves toward reimperialization. This scenario has been dubbed "Weimar Russia," i.e., the possibility that, embittered by its economic and political troubles and humiliations, Russia may attempt to recover its past glory by turning to ultranationalist policies, particularly the reincorporation—or hegemony over—of part or all of the old "internal" and even external empire.

Russian statements indicate a strong preference for reincorporation of the so-called "near abroad" (the states on the territory of the former Soviet Union) and increased Russian international assertiveness. Relations between Russia and the United States and its European allies have worsened because of Russian brutalities in Chechnya, the sale of reactors to Iran, Russian support for lifting the embargo on Iraq, and Russian opposition to NATO expansion. They might worsen even more when NATO expands eastward. Russia might respond to NATO expansion by improving relations with Iran, China, or India. It might also seek to reincorporate Ukraine and the Baltic states in the Russian security zone. The challenge for the West is how to expand NATO—without precipitating further negative changes in Russian domestic and foreign policies. However, a Russian expansion into Ukraine and the Baltic states is likely to produce a new containment strategy with major implications for U.S. military planning.

In Asia, the People's Republic of China might, over the long term and perhaps sooner than Russia, emerge as a global rival by first dominating East Asia. China's economic dynamism is now being reflected in its military development. Chinese investment in its military has grown significantly over the past several years. Militarily, China has been increasing its power-projection capability—both naval and air—in part by purchasing advanced equipment from Russia. China has also been importing Russian military scientists to help increase domestic production of sophisticated equipment.

However, China faces significant political uncertainties in its domestic politics, including a possible succession crisis on the death of Deng Xiaoping and the centrifugal tendencies unleashed by differential economic growth among the provinces. Indeed, Chinese weakness, not excluding a possible civil war that could disrupt economic prosperity and create refugee flows, may cause significant problems for its neighbors and the world community.

But China is an ambitious power. Among the major powers, China appears more dissatisfied with the status quo than the others. Beyond Hong Kong and Macao, which will be ceded to China by the end of the century, it claims sovereignty over substantial territories that it does not now control—such as Taiwan, the Spratly Islands and the South China Sea generally, and the Senkaku Islands between China and Japan. While China has abandoned communism as a global ideology and seems to have accepted the economic imperative of the global economy, it is, geopolitically, still seeking its "rightful" place in the world. How will China define its role as its power grows beyond its territorial interests? China appears to be seeking eventual regional predominance, a prospect opposed by Japan, Russia, India, Indonesia, and other regional powers. Even without regional domination, it might become interested in becoming the leader of an anti-U.S. coalition—based on a rejection of U.S. leadership generally or as it is expressed in such policies as nonproliferation and human rights. This is evident in its assistance to Pakistani and Iranian nuclear programs. Chinese writings on strategy and international security express hostility to U.S. predominance and imply the need to balance it. But China recognizes the importance of the United States— as a market for Chinese goods and a source for technical training and technology. Without U.S. help, China is less likely to achieve its economic and military objectives.

China, however, is decades away from becoming a serious global rival either by itself or in coalition with others. This provides the United States with ample strategic warning. For the near term, economic considerations are likely to be dominant in Chinese calculations. Chinese economic success confronts the United States with a dilemma. This success could increase the Chinese potential for becoming a global rival. On the other hand, it might produce democratization and decentralization and a cooperative China.

Even in its current state, China could, by itself or as the leader of a coalition of renegade states, increase the global proliferation problem in key regions such as the Persian Gulf and Northeast Asia. To maintain its global leadership role, it is not in the U.S. interest to cut off ties with China or to isolate it. A leadership strategy requires that the United States should limit the transfer of technologies, which can have important military implications. It also means ensuring that Chinese neighbors, such as Taiwan and the Association of Southeast Asian Nations (ASEAN) states, have the means to defend themselves. Because of their increasing economic importance and as a hedge against an expansionist China, the United States should strengthen its ties with ASEAN states. The U.S. military would have to strengthen ties such as access, joint training, and military-military contacts with the ASEAN countries. Cooperation between the United States, Japan, and other regional powers should discourage possible Chinese expansionism. But should the Chinese become expansionist, such cooperation places the United States and its allies in a strong position to contain that expansion.

A global leadership role has many implications for the U.S. military, including the Air Force. To shape the future, build U.S. influence, and deter the rise of a rival, the U.S. armed forces would need to have the following broad capabilities:

First, sustain nuclear deterrence against possible attack by Russia or China.

Second, fight and win major regional conflicts.

Third, deter and defend against the use of weapons of mass destruction (WDMs) and missiles in regional conflicts. Future regional conflicts are likely to involve the use of WMDs and missiles by hostile forces. Therefore, the U.S. armed forces need to acquire increased capability to deter, prevent, and defend against the use of biological, chemical, and nuclear weapons in major conflicts in critical regions. The regional deterrence requirements might well be different from those appropriate for the Soviet Union during the Cold War because of the character and motivations of different regional powers. The U.S. ability to prevent use and defend against use is currently very limited. In the near term, therefore, to deter use of WMDs against its forces and allies, the United States may have to threaten nuclear retaliation.

Fourth, improve the capability to fight LRCs—internal conflicts, small wars, humanitarian relief, peacekeeping or peacemaking, punitive strikes, restoring civil order, evacuation of Americans, providing security zones, and monitoring and enforcement of sanctions

(Builder, 1994, pp. 225–237; Kassing, 1994; Lempert et al., 1992). Even in a global leadership framework, the United States can be selective in its military involvement around the world. Nevertheless, some LRCs may occur in areas of vital importance to the United States—e.g., in Mexico or Saudi Arabia—and others might so challenge American values as practically to require U.S. military involvement. The United States might also consider participating with allies in some LRCs because of a desire to either extend the zone of peace or prevent chaos from spreading to critical regions in ways that would undermine the zone of peace and prosperity. At present, LRCs are treated as lesser included cases, much as some thought about regional conflicts during the Cold War. The United States "underestimated and misestimated the MRC requirements during the Cold War." (Lewis, 1994, p. 103.) Now, it would be a mistake to treat LRCs the same way, especially since future U.S. forces will be much smaller than in the past and will provide far less slack. Even small LRCs can impose substantial and disproportionate demands on the support elements of U.S. forces, such as the Airborne Warning and Command System, Suppression of Enemy Air Defense, airlift, and communications. The MRC-driven force structure must be modified as well in connection with the training and organization of U.S. forces to improve their capabilities for LRCs.

Fifth, retain a mobilization base to "reconstitute" additional capability in a timely fashion if things go badly in any major region. Without such a capability, the United States is unlikely to be able to take timely action, given the strategic warning that is likely to be available.

Sixth, change the planning approach. To discourage the rise of another global rival or to be in a strong position to deal with the problem should one arise, the current Korea-and-Gulf-focused approach, plus increased ability for LRC and counterproliferation operations, is inadequate for force sizing. Over time, North Korea will probably disappear, and other larger threats could emerge. As an alternative, the United States should size its forces by requiring them to have the capability to defeat nearly simultaneously the most plausible military challenges to critical American interests that might be created by the two next most powerful military forces that are not allied with the United States. Such a force backed by *rapid global mobility and presence* should allow the United States to protect its interests in Asia, Europe, and the Persian Gulf. Such a force-sizing principle does not mean that U.S. forces have to be numerically as large as the combined forces of these two powers. It means that U.S. forces should be capable of defeating them in relatively specific areas and scenarios of great importance to the United States. Such an approach would give the United States a flexible global capability for substantial operations.[2]

Seventh, control space and dominate the development of new military technologies and the concepts for their use in such areas as information warfare. Therefore, the United States should give higher priority to research on new technologies, new concepts of operation, and changes in organization, with the aim of U.S. dominance in the military revolution that may be emerging. The Gulf War gave a glimpse of what is likely to come. The character of warfare will change because of advances in military technology, where the United States has the lead, including commensurable concepts of operation and organization structures. The challenge is to sustain this lead. However, given the current U.S. lead in military power, there is some danger of U.S. complacency. Those who are seeking to be rivals to the United States are likely to be very motivated to explore new technologies and how to use them against it. A determined nation making the right choices, even one with a much smaller economy, could pose an enormous challenge, making more traditional U.S. military methods less relevant by comparison.

Whether the United States retains its global leadership position will depend in large part on what happens in the United States. One factor that will be key will be the state of the U.S. economy. The United States is unlikely to preserve its military and technological dominance if the U.S. economy declines seriously or if the balance of economic power shifts decisively to another country. In such an environment, the domestic economic and political base for global leadership would diminish, and the United States would probably incrementally withdraw from the world. As the United States weakened, others would try to fill the vacuum. The world is likely then to become multipolar. Therefore, leadership requires a strong U.S. economy.

---

[2]Some of the points here regarding military challenges of the new era are also discussed in Davis (1994b).

Another key variable is the attitude of the American people. Public opinion polls indicate that the American people are focused on domestic concerns. However, according to polls on American public attitude, Americans appear to support active U.S. involvement in world affairs. At the same time, 84 percent believe that we should pay less attention to international problems and concentrate on problems here in the United States. A majority of Americans support peace "through military strength." (Times Mirror Center for the People and the Press, 1994, p. 37.)

Whether the public would in fact support a global strategy—as outlined here—is not known. The public might well support it if (a) it were presented to them by the president and supported by the senior members of both the Democratic and Republican parties and (b) if the costs and benefits of such a strategy and some alternatives were debated and understood. A global leadership strategy will entail costs—a greater defense effort in the near term than would be the case if some other grand strategy were adopted—but those costs have to be compared with the potential risks of alternatives. At present, the burden imposed by U.S. defense efforts, approximately 4 percent of GNP, is lower than at any time since before the Korean War.

Another factor, which will determine whether a leadership role can be sustained, is the degree of U.S. involvement in conflicts around the world. Overextension is a mistake that some of the big powers in the past have made (Kennedy, 1987). Such a development can occur if the United States is not judicious in its use of force and gets involved in protracted conflicts in various regions, sapping its energies and undermining support for its global role. In a leadership role, U.S. vital interests would be engaged in critical regions where it would have to be prepared to use force if other means fail. And when the United States uses force, its preference should be to have its allies and friends go in with it.

When it comes to lesser interests, the United States could rely on nonmilitary options, especially if the military costs do not warrant the stakes involved. There are other options here: arming and training the victims of aggression, providing technical assistance and logistic support for peacekeeping by the UN, regional organizations, or powers; economic instruments—sanctions and positive incentives—and, of course, diplomacy.

Within these constraints, it is in the U.S. interest and the interests of other members that the zone of peace ultimately encompass the whole world. The reason for favoring such an evolution is that prosperous democracies are more likely to cooperate with the United States and are less likely to threaten U.S. interests. Unfortunately, this is not a near-term proposition. Many regions and states are not ready. The United States and its allies should seek to expand the zone selectively and help others prepare for membership.

The most important step that the United States and the other prosperous democracies can take is to assist others in adopting the economic strategies that have worked in North America, Western Europe, and East Asia and that are being successfully implemented in parts of Latin America and elsewhere in Asia. Economic development and education are the most effective instruments for solving the problems of the nations in the zone of conflict.

The members of the "zone of peace" have a common interest in the stability of Europe, North America, East Asia, and the Persian Gulf. Japan, for example, imports oil from the Gulf and exports to and invests in the other critical regions. The same is true of Europe. The U.S. global leadership role benefits the United States and these other members. There is a danger that the other members of the zone of peace will not do their fair share and perpetuate free-ridership. This was a problem during the Cold War, and it is unlikely to go away. It could become a bigger problem if, because of the absence of the Soviet threat and the lack of a common objective, burden-sharing declines. It is clearly an important political issue in the United States. The nation does face a dilemma: As long as it is able and willing to protect common interests, others might be happy to rely on it, thereby keeping political opposition under control, accepting no risk for their youth, and continuing to focus on their own economies. A global leadership role requires that the United States should be willing to bear a heavier military burden than its allies. However, political realities in the United States require that this disproportion not be excessive. A balance needs to be struck, and a formula has to be found to balance each country's contribution of "blood and treasure" in protection of common interests. For the long term, one possible solution is to institutionalize burden-sharing among the G-7 nations for the security of critical regions, including sharing the financial costs of military operations. Questions of out-of-area responsibility are important in peacetime, both on a day-to-day basis and in times of crisis and war. Effective burden-sharing will also place some constraints on U.S. policy. It will mean that the United States would have to pay greater attention to their views and concerns and be willing to put American lives at risk to protect common interests.

CONCLUSION

The United States could choose one of the three ideal-type grand strategies discussed here. Alternatively, it could decide not to choose and instead react to developments in the world on a case by case basis, judging the importance of specific events by the conditions— both domestic and international—at the time. Besides losing the opportunity to shape the future, the latter option poses a particular problem for the U.S. military services. Unless they have the resources to prepare for the entire range of plausible conflicts in the world—an unlikely prospect—they would be in a less favorable position to plan for the long term.

REFERENCES

Builder, Carl, "Nontraditional Military Missions," in Charles F. Hermann, ed., *American Defense Annual,* 1994 ed., New York: Lexington Books, 1994, pp. 225–237.

Davis, Paul K., "Protecting the Great Transition," in Davis (1994b), 1994a, pp. 135–164.

_____, ed., *New Challenges for Defense Planning,* Santa Monica, Calif.: RAND, M-400-RC, 1994b.

Huntington, Samuel, "The Clash of Civilization," *Foreign Affairs,* Vol. 72, No. 3, 1992.

"The Initial Draft of the Defense Department's Planning Guidance," *New York Times,* March 8, 1992, p. 1.

Kassing, David, *Transporting the Army for Operation Restore Hope*, Santa Monica, Calif.: RAND, MR-384-A, 1994.

Kennedy, Paul, *The Rise and Fall of the Great Powers*, New York: Random House, 1987.

_____, *Preparing for the Twenty-First Century*, New York: Random House, 1993.

Kissinger, Henry, *Diplomacy*, New York: Simon and Schuster, 1994.

Krauthammer, Charles, "The Unipolar Moment," *Foreign Affairs: America and the World*, Vol. 70, No. 1, 1990–1991.

Lempert, R., D. Lewis, B. Wolf, and R. Bitzinger, *Air Force Noncombat Operations: Lessons From the Past, Thoughts for the Future*, Santa Monica, Calif.: RAND, N-3519-A, 1992.

Levin, Norman D., ed., *Prisms and Policy: U.S. Security Strategy After the Cold War*, Santa Monica, Calif.: RAND, MR-365-A, 1994.

Lewis, Kevin, "The Discipline Gap and Other Reasons for Humility and Realism in Defense Planning," in Davis (1994b), pp. 101–132.

"Strategy and the Internationalists: Three Views," *RAND Research Review*, Vol. XVIII, No. 1, Summer 1994, pp. 2–5.

Times Mirror Center for the People and the Press, *The People, The Press & Politics*, Washington D.C., September 21, 1994.

Part Two

Chapter Two

WESTERN EUROPE

Ronald D. Asmus

Six years following the collapse of the Berlin Wall, a new geopolitical drama is unfolding across Europe's landscape. In the immediate wake of communism's demise, Europe appeared to be on the threshold of a new era. Eastern Europe was liberated, Germany was unified, and the troops of the former Soviet Union were being withdrawn some 1,000 km eastward. The United States and Western Europe embraced the vision of a new Europe whole and free built upon a strengthened European Union (EU) and a transformed Atlantic Alliance. History may not have ended, but the dream of the generation of European leaders who emerged from the Second World War determined to end once and for all internecine European conflict and to build a new Europe based on integration and cooperative security appeared to be within reach.

By 1995, however, the initial euphoria sparked by communism's collapse and the end of the Cold War had faded. War and genocide in the former Yugoslavia, growing uncertainty over Russia's future course, a slowdown of reform efforts in Eastern Europe, the collapse of momentum in European integration, and growing uncertainty over the future U.S. role and commitment—all these factors underscore Europe's new instability. As Europe looks eastward, the greatest instability and the most dramatic scenes on this new political stage are currently taking place in the Balkans, as well as in Russia and the newly independent states of the former USSR. The great fear is, of course, that the kind of destructive nationalism that has been unleashed there will spread to Eastern Europe as well.

As Europe looks south, there are also signs of growing instability. If one assumes a broad definition of the Mediterranean region, ranging from the Maghreb to the Middle East and extending into the Caucasus, this strategic space is one of the least stable and most likely to produce conflict and war. The potent combination of radical Islam, trends toward proliferation, and the uncertainty of traditional pro-Western pillars, such as Turkey, produced enormous strategic uncertainty. Often considered Europe's strategic backwater during the Cold War, the Mediterranean has in many ways emerged as Europe's new front line as the West awakens to the new strategic challenges that may be beginning in this region.

While no longer divided, Europe also no longer seems as stable as it initially did in the wake of the end of the Cold War. The vision of a whole and free Europe seems increasingly elusive. The prospect of new instability in the East and the South has also raised the danger of spillover effects destabilizing the traditionally stable democracies of Western European countries. This has underscored how old Cold War distinctions between in-area and out-of-area crises have become politically archaic and how the destinies of Western, Eastern, and Southern Europe are intertwined. Central European, North African, and Middle Eastern security concerns are intersecting in new ways, blurring old strategic distinctions that have guided past policy. Europe faces the dilemma of trying to export stability into these regions or run the risk of importing instability.

At the same time, the West's relations with Russia are also increasingly strained. Moscow has reacted to plans by the EU and the North Atlantic Treaty Organization (NATO) to enlarge eastward to stabilize the continent's eastern half with escalating rhetoric and warnings that such moves will produce a new Cold War in Europe. The West's desire to integrate the two halves of a continent divided during the Cold War has clashed with Moscow's reluctance to abandon its former influence in this region and Russian fears about being "locked out" of the new Europe. As a result, the West and Russia are increasingly engaged in a new and more subtle form of geopolitical competition over the future destiny of the so-called "lands in between" Germany and Russia in Eastern Europe. More broadly, the entire edifice of the West's relations with Russia may increasingly be in jeopardy. Many key arms control agreements from the Cold War era, such as the Conventional Forces in Europe agreement, the Strategic

Arms Reduction Talks, and the Antiballistic Missile treaty, codified an old political and strategic order that is gone.

What are the implications of these trends for U.S. interests? In the immediate aftermath of communism's collapse, there was a widespread sense that the issue of Europe, as a major U.S. national security concern, was finally resolved. Some observers even went so far as to deem Europe irrelevant as a major national security issue. The U.S. decision to maintain nearly 100,000 troops in Europe notwithstanding, Europe ceased to play a major role in U.S. defense planning priorities, as underscored by the Clinton administration's Bottom-Up Review.

In recent years, however, a growing chorus of voices has warned that European security remains more fragile and vulnerable than often realized and that U.S. engagement in Europe is vital if the continent is to remain stable and if vital U.S. interests are to be protected. The growing debate over Europe's new instability also reflects a growing intellectual and political dispute with important policy ramifications. As Europe's new strategic landscape unfolds, the United States must decide when and where vital U.S. interests are at stake and how it wants key European security institutions to evolve to meet these new challenges.

The purpose of this chapter is to step back and identify broader security trends in Western Europe that are of strategic significance for the United States. The chapter addresses four issues. First, it assesses Europe's new strategic landscape. Second, it briefly examines how the largest and most important Western European countries and key allies of the United States—Germany, France, and the United Kingdom—have responded to these changes. Third, it examines the factors that have led to the collapse of the Maastricht vision of the EU and the new debate unfolding within the EU over its future in the run-up to the 1996 Inter-Governmental Conference (IGC). Fourth, it looks at the unfolding debate over NATO's future, how the debate reflects the broader issue of the future U.S. role in Europe, and what that role could mean for future U.S. defense planning.

EUROPE'S NEW STRATEGIC LANDSCAPE

The end of the Cold War wiped away the strategic distinction between Europe's center and its periphery that had come to dominate much of Western strategic thinking in the postwar period. The locus of potential conflict in Europe has shifted from the heart of the continent, reflected in the Cold War standoff at the former inter-German border, to Western Europe's periphery in the East and the South. The revolutions of 1989 not only led to the collapse of communism and the subsequent unraveling of the USSR, but have also unleashed a new set of dynamics that have overturned the orders of Yalta and Versailles established in the wake of two world wars this century. German unification overturned the former, and the subsequent unraveling of the USSR, Yugoslavia, and Czechoslovakia has largely undone the latter.

However, what many viewed as Western Europe's "periphery" during the Cold War is in reality central to future European security. Europe's new instability now falls along two geographic arcs. The first is the eastern arc, running from Northern Europe south between Germany and Russia through the Balkans. The second is the southern arc, running through North Africa and the Mediterranean into Turkey, the Middle East, and Southwest Asia. While seemingly located safely on Western Europe's outer edges, developments in these regions are crucial for the continent's overall stability.

Europe's eastern arc of crisis encompasses one of the most geopolitically sensitive regions of the continent. It is here that two World Wars, as well as the Cold War, started. The collapse of Soviet power and the old Cold War system left in its wake a strategic vacuum between a unified Germany integrated into the EU and NATO and a weakened and truncated Russia. Europeans remain concerned about the rise of nationalism and the possible failure of democracy and reform across this region. Such instability could unleash political, economic, and nationalist forces that would undercut democracy in Western and Central Europe as well. Lurking in the background is the fear of the same process leading to a shift toward the right and imperial restoration in Russia.

At the same time, Europeans are also increasingly aware of a new set of strategic challenges that could emerge along the southern arc of crisis as well. If the Mediterranean is defined in the broad sense as encompassing the region extending from Gibraltar to the Black Sea, the area is complete with crises, both potential and real. In addition to the current war in the former Yugoslavia, the United States and its allies also face the prospect of growing political instability in Northern Africa. Indeed, Algeria could well prove to be the next crisis that could require a coordinated Western response. In addition, Turkey's future strategic orientation is also becoming increasingly less certain. Looking further into the East, the West faces the prospect of geopolitical turbulence in the Transcaucasus that could involve Iran, as well as Turkey and Russia.

The Mediterranean is much more than a body of water. It defines an extraordinarily diverse region. It forms a space that is more diverse and disparate than unified, with its various subregions tied together by the common fact that they all border on a shared body of water. Often considered Europe's strategic backwater during the Cold War, the Mediterranean has in many ways emerged as Europe's new front line as the West confronts the strategic challenges of the post–Cold War era. It is here that some of the most likely candidates for future European security crises are located, yet the West is in many ways the least well prepared to address crises in this region. These crises range from instability in North Africa, unrest in and radicalization of the Muslim world, and refugee flows to the fact that, by the end of the 1990s, it is likely that all major southern European capitals will be within range of ballistic missiles based in North Africa and the Levant. Although in recent years the debate over European security has been dominated by the war in Bosnia and instability in the East, policymakers in Madrid, Paris, and Rome are waking up to a possible strategic nightmare on their southern doorstep.

The differences and symmetries among the challenges along Europe's eastern and southern arcs are obvious. History, geography, and culture make enormous differences. While the EU and NATO debate over when and how to expand into the East, such issues are not on the European agenda vis-à-vis the South. Several less apparent similarities are potentially crucial for policymakers. Instabilities in the East and the South are part of a broader set of strategic challenges beyond Western Europe's immediate borders. They are also part of a common strategic problem facing these countries—the need to export stability or run the risk of importing instability.

Although the nature and magnitude of the challenges along the two arcs may differ, they reflect the fact that Cold War distinctions have become obsolete in the Europe of the 1990s. Just as the definitions of Western and Eastern Europe are increasingly blurred in security terms, so is the traditional distinction between Mediterranean and Middle Eastern security as the strategic concerns of Southern European countries increasingly extend further into the Muslim world and the Middle East. And just as Western Europe has been compelled to extend its strategic horizon eastward, after the collapse of communism, a different but significant set of forces is increasingly drawing Western Europe's strategic attention southward as well.

A second similarity exists regarding building a Western consensus on how to respond to these challenges. If the EU and NATO are to reach consensus and be able to act effectively in addressing these challenges, they will have to find a balanced approach that simultaneously addresses divergent concerns among Western Europe's key actors in the East and the South. While one group of countries led by Germany remains primarily concerned with the East, France, Italy, Spain, and Turkey are first and foremost concerned about national security concerns in the South. While Germany seeks to enlist the support of its key allies in addressing instability in the East, others express concern that Germany's agenda could crowd out their own concerns. The issue is whether key European security institutions, such as the Atlantic Alliance and the EU, will be able to address both sets of issues simultaneously—or run the risk that growing divisions among member states will render them unable to address either.

The prospect of new instability in the East and the South has underscored how old Cold War distinctions between in-area and out-of-area crises have become politically archaic. Just as the definitions of Western and Eastern Europe are increasingly blurred in security terms, so is the traditional distinction between Mediterranean and Middle Eastern security as the strategic concerns of Southern European countries increasingly extend further into the Muslim world and the Middle East. And just as Western Europe has been compelled to extend its strategic horizon eastward, after the collapse of communism, a different but significant set of forces is increasingly drawing Western Europe's strategic attention southward as well. Europe faces the dilemma of trying to export stability into these regions or run the risk of importing instability. How to deal with this new instability in and around Europe is increasingly seen as *the* new European security problem (Freedman, 1994).

The core and unresolved issue remains the age-old question: What is Europe? And there are the attendant issues: Who decides and what are the criteria? Where does it end? What are the consequences of line-drawing? What kind of strategic relationship will emerge with those countries on the periphery of the new Europe, hostile or friendly? The new democracies in East-Central Europe, such as newly independent Ukraine, wonder whether, in Western European eyes, they will become bridges, buffers, or barriers as the EU and NATO define their outer limits. In the South, there is worry of a new Cold War as Europe defines its post–Cold War relationship with Northern Africa and the Muslim world.

In short, the political changes triggered by the collapse of communism are transforming Europe's strategic landscape, producing a new and often complex geopolitical tapestry with many different shadings and themes. Former German Foreign Minister Hans-Dietrich Genscher's comment during the fall of 1989, the heady days of the collapse of communism, that "nothing will ever be the same" has turned out to be true in more ways than many imagined. While the single overriding threat of the Cold War has been removed, there are many different potential sources of instability that could unravel corners or even large parts of this new tapestry. After all, Europe before the Cold War was hardly an island of stability.

Perhaps the key factor that will determine Europe's destiny is the struggle between the forces of integration and disintegration currently being played out in different ways in and around Europe. Will Europe succeed in extending the successful experience of Western political and economic integration and collective security eastward and southward? Can the positive impact of Western integration overcome residual nationalism and old patterns of geopolitical rivalry and conflict in Europe's eastern half? Or will the forces of disintegration prove too strong? Does Europe again run the risk of importing new instability and possibly fragmenting?

THE NEW POLITICAL LANDSCAPE: GERMANY, FRANCE, AND THE UNITED KINGDOM

The country most directly impacted by these changes, and whose future stability and orientation are crucial for the continent's future, is, of course, Germany. Germany was in many ways the great beneficiary of the end of the Cold War. The collapse of communism paved the way for German unification under Western auspices and the regaining of full German sovereignty. With the completion of the withdrawal of the troops of the former Soviet Union, Russian military power has been moved some 1,000 km eastward. Germany is now surrounded by smaller, weaker, and friendly states—a constellation that gives the country enhanced strategic depth and greater international leeway than at any time in the postwar period.

At the same time, Germany's geography means that it remains vulnerable to the spread of Europe's new instabilities. The challenges of the rapid reconstruction of the new eastern states have drained the country's financial coffers and strained the country's political fabric. Although the country's party system has not yet been hit with the shocks that toppled governments in France and Italy, growing talk of a possible political realignment of the postwar party system has surfaced, especially if the ruling coalition should stumble. Economically, the burdens of unification have exacerbated the country's declining competitiveness problem, have helped fuel unemployment levels not seen since the early 1950s, and have led many to question the sustainability of the "German model." Some observers already speak of the emergence of a new Berlin Republic (see Hamilton, 1994).

Perhaps most important from the perspective of American national security interests, Germany's strategic calculus is changing in the wake of the geopolitical restructuring of Europe. Several years ago, German newspapers decreed the collapse of "Genscherism"—a reference not only to the resignation of the long-standing German Foreign Minister but a reflection of the fact that the foreign policy that served the Federal Republic so well during the 1970s and 1980s was increasingly passé in the 1990s. Unification has again thrust the country into the continent's center, or *Mittellage*, and the country's foreign policy and strategic culture are being reshaped by these new geopolitical circumstances.

What is important for our purposes is to understand Germany's new emerging strategic agenda. The need to stabilize Germany's eastern flank is rapidly becoming the number one security concern for the German political class. There is a clear consensus that the country's eastern border cannot remain the eastern edge of either the EU or NATO and that Germany must seek to expand the institutions of the West to secure the country's vulnerable geopolitical position and to prevent the reemergence of old dynamics and rivalries that have so often been the source of conflict in Central Europe in the past. This has little to do with some mythical *Drang nach Osten*, but instead results from a *Zwang nach Osten*—or the imperative to become more involved in the East to project democracy and stability and to prevent the emergence of instability on the eastern border from destabilizing Germany itself. Germany's response has been to turn to its key allies and those Western institutions that have guaranteed German security in the postwar period and to transform them to meet these new vulnerabilities.

In light of the magnitude of the challenge of rebuilding eastern Germany and stabilizing Eastern Europe, German leaders are almost desperate to enlist the help and cooperation of their Western allies in integrating the new democracies. At the same time, Germans are increasingly realizing that such institutions will only move in this direction, if at all, if Germany emerges from the geopolitical niche it occupied during the Cold War and becomes a more active player in determining the future gestalt of Europe. Nowhere, however, is Germany's new external agenda more apparent than in the importance attached to the new Eastern Question and the need to develop a new German *Ostpolitik* through both the EU and NATO.

German leaders have repeatedly underscored their commitment to a broadening of the EU to the East. In German eyes, EU expansion to Austria, Sweden, and Finland in early 1995 was but a prelude to the eventual incorporation of some ten additional countries. These include the six Eastern European countries already singled out by the EU (the Visegrad 4 plus Bulgaria and Romania), the three Baltic states, and Slovenia. While Bonn's top priority is a rapid expansion to the Visegrad countries, the longer-term German agenda envisions an EU with nearly 25 to 30 members. Such an expansion is also an enormously powerful geopolitical move, because it would also pave the way to new security relationships and, ultimately, either implicit or explicit security guarantees to these countries. From a German viewpoint, the EU would thus cover all those areas in which Germans consider themselves to have core interests.

At the same time, Germany has sought to underscore that its commitment to the broadening of the EU will not come at the expense of the deepening of integration. Many in the German political class, especially Chancellor Helmut Kohl, are among the residual true believers in the original goals of European integration, still embracing the old federalist vision at a time when it is clearly out of vogue in much of Europe. The reason is simple. For Kohl, as well as many other German leaders, deepening European integration is the guarantee against a renationalization of both European and German politics at a time when nationalism in other parts of Europe is on the rise. Germany is perhaps the key country pushing the EU to expand, and its leaders know they must also work for reforms that will render an EU of that size viable and effective.

It is also with regard to NATO, however, that the shift in German strategy is evident. It was German Defense Minister Volker Rühe who publicly launched the NATO expansion debate with his International Institute for Strategic Studies (IISS) speech in the spring of 1993, arguing that NATO's political and military shield is needed to stabilize fragile democracies in East-Central Europe (Rühe, 1994). German support for NATO expansion is rooted in a combination of history, culture, and geopolitics. More than any other Western country, Germany understands the dangers of the new political and security vacuum left behind in the wake of the collapse of communism. Bonn's commitment to a rapid expansion of the EU simultaneously pushes it in the direction of a rapid expansion of NATO. By the spring of 1995, a solid consensus had formed in the German government on these issues, underscored when Chancellor Helmut Kohl announced in Warsaw that Bonn planned to push for both EU and NATO enlargement by the year 2000 (Kohl, 1995).

These shifts in Germany's strategic priorities have been reflected in changing German defense planning priorities and the restructuring of the Bundeswehr, as well as in shifting trends in German public opinion. Since the end of the Cold War, Germany has moved further and faster in shedding many of the constraints on the use of the German armed forces than many observers had initially anticipated.

The legal question about the use of the German armed forces in missions beyond territorial defense has been removed and a number of key political constraints have been relaxed. The German armed forces have embarked on a reorientation and restructuring process that has been rightly termed the second birth of the Bundeswehr. When fully implemented later this decade, it is designed to give Bonn the capability to project military forces of up to two divisions and to operate as a key coalition partner well beyond Germany's borders.[1]

Shifts in German public opinion reinforce this trend as well. Whereas the participation of German armed forces in so-called "out of area" operations led to an enormous political and constitutional battle that was only clarified with the German Constitutional Court ruling of July 1994, the issue of NATO expansion has received much broader support across the political spectrum and in the German public. A multiyear study on changing German public opinion on such strategic issues sponsored by RAND and the Friedrich Naumann Foundation has found, for example, that the German public has made the conceptual leap to a new NATO embracing new missions and that solid majorities support both NATO enlargement and missions beyond NATO's borders.[2]

Germany's attempts to emerge from the geopolitical niche it occupied during the Cold War and to define a new role and agenda in Europe have been matched by France's attempt to redefine its rank and role in Europe. The great paradox for France is that its jubilation at seeing the Berlin Wall crumble was mixed with apprehension over German unification and a sense of loss, the loss of the privileged position Paris had enjoyed throughout much of the postwar period. Geography and Europe's division had allowed France under de Gaulle to carve out a special position of influence in both transatlantic and European affairs.

With the collapse of communism and the end of East-West confrontation in Europe, many of the factors that had increased France's influence and prestige were now devalued. Nowhere was this more clear than in France's relations with Germany, in which the end of

---

[1]For further details see Asmus (1995).

[2]For further details see Asmus (1994).

the latter's division meant the loss of French influence. The tables were turning in the Franco-German relationship, with Germany gradually emerging as the senior economic partner and one that is also less dependent upon France in security terms. At the same time, the need to deal with the enormous challenges in the East meant that Europe's political attention was increasingly drawn in that direction—and away from the French vision of a tightly integrated Western Europe led by France. In its place, a much looser and broader Europe, in which Germany is likely to emerge as the single most-important power, seems to be the wave of the future.

Against this background, it is hardly surprising that France has been forced to redefine its priorities and policies, and that this process has led to considerable internal strain. Perhaps the first victim was French European policy, in which new fault lines in French politics soon emerged. A gap has emerged between those who still see the EU as the best means to preserve French and limit German influence—but from a new, different, and in many ways weaker position—and those who fear that Germany will dominate the Union and constrain French independence.[3] It is noteworthy that the French government's main argument in favor of European integration in the run-up to the referendum on the Maastricht Treaty was the need to contain and constrain a unified Germany.

The first open sign of a significant shift in French sentiments, however, came to the fore in that narrowly won referendum on Maastricht. Despite defeat, critics and skeptics of European integration seemed further emboldened. Sensing growing anxiety about the erosion of national sovereignty and French national independence, French politicians across the board have backed away from notions of "federalism" or the vision of a United States of Europe. The surprise decision by Jacques Delors in December 1994 to spurn a presidential race many thought he was destined to win only seemed to concede just how the political winds in Paris had changed, especially when Delors justified his move by admitting in public that, even if elected, he no longer saw a majority in France prepared to implement his vision of a unified Europe.

---

[3]See the excellent essay by Hoffman (1994), pp. 1–25.

French policy has sought to maintain a careful balancing act. Recognizing the centrality of the Franco-German relationship and the importance to Bonn of Germany's new eastern agenda, French leaders have found themselves with little choice but to support Germany's interests in the eastern expansion of the EU and, to a lesser degree, of NATO as well. At the same time, French leaders have underscored the need to ensure, for example, that the EU's northern and eastern expansions do not marginalize it. Paris has therefore both continued to pursue Franco-German partnership and sought to build its ties with the United Kingdom, as well as key Mediterranean countries, to maintain its influence in an expanding European Union (Balladur, 1994). While officially rejecting the notion of a new "hard core" within the EU, Foreign Minister Alain Juppé and others have talked of a series of concentric circles within the EU—with an inner circle building a monetary and military union; within a second circle in which members would cooperate in trade, as well as foreign and security policy; and with a much looser outer circle embracing all of Europe, which would cooperate still more loosely on matters of diplomacy and trade (Juppé, 1995).

At the same time, the French have also been among the first to realize that Europe no longer has a security system capable of confronting the potential for new instability in the East and in the South by itself. It is here that the evolution in French thinking has been the most interesting, at times ambivalent and still unresolved. France has clear national interests along both arcs of crisis in the new Europe. Its close ties to Germany mean that the latter's concerns about instability on its eastern flanks have also become those of France and the EU. French policy has, in turn, increasingly discovered Europe's new Eastern Question in an attempt to gain a voice of codetermination with Germany in setting Western Europe's new *Ostpolitik*.[4]

Yet, France is even more directly threatened by instability in Northern Africa and the Muslim world and the prospect of proliferation. For this reason, Paris has taken the lead in trying to launch a new Euro-Mediterranean partnership within the EU, working with Spain and Italy to ensure that Germany's desire to bring the EU's

---

[4]For further details on French Eastern policy, see Larrabee (1994), pp. 76–78.

resources to stabilize the East are matched by an equal effort by the EU to extend stability in the South. Similarly, within the Atlantic Alliance, France has been a voice insisting that NATO's efforts at political outreach to the East be matched by attempts to build new partnerships in the South.

In both cases, France needs strategic cooperation with the United States. As Pierre Lellouche wrote in *Foreign Affairs* in the spring of 1993:

> Once again, Europe is characterized by a pivotal and strong Germany, a backward and unstable Russia, and a large number of small, weak states. And again, France and Great Britain are incapable by themselves of balancing German power or checking Russian instability, let alone restructuring the entire European order around a Franco-British axis.[5]

The result has been the collapse of the Gaullist consensus. If the end of the Cold War marked the end of Genscherism in Germany, its victim in France was the battered legacy of de Gaulle. One of the great paradoxes of current French politics was the election of Jacques Chirac, a political protégé of de Gaulle, to the presidency in the spring of 1995, precisely when many of the assumptions underlying de Gaulle's European strategy had crumbled. Many French commentators nonetheless pointed out that if anyone on the current French political scene was capable of breaking out and defying and redefining the contours of traditional Gaullist and French thinking for the post–Cold War era on these issues, Jacques Chirac was that figure.

One result of this shift has been a French rethinking of the role of the United States in European security and a guarded French attempt to seek rapprochement with Washington. With the drawdown in U.S. forces in Europe, French concerns have shifted from worrying about predominant American influence to worrying about too little American engagement. French officials insist that they will not give up their commitment to building an integrated Europe, yet acknowledge that the pre-1989 French vision of a tightly integrated Western Europe led by France is part of the past. While insisting that France

---

[5]See Lellouche (1993), pp. 122–131.

does not want to return to the old NATO—for, in their eyes, that would simply legitimize the status quo and prevent needed major adjustments—French officials underscore that Paris is willing to build a new NATO. The issue of building Europe "against" the Americans has been transformed into a search for ways to try to build Europe "with" the Americans, while continuing to maximize French influence at the margin, a shift that has been made easier by the Clinton administration's clear support for European integration (see Clinton, 1994c).

The impact of the end of the Cold War on the United Kingdom has, at first glance, been less dramatic because of the dictates of geography. At the same time, London must define its interests in a changing European landscape, above all in the East and South. It must also decide where and how to bring its influence to bear in constructing a new European order. To be sure, London, too, is concerned with containing a unified Germany whose arrival British leaders greeted with barely concealed dismay (see Thatcher, 1993). But a tightly integrated EU appears no less capable of performing such a function to British leaders than it does to anti-Maastricht French politicians, such as Séguin or Chevènement. Both would like a much looser Europe, extending much further East, both to stabilize the East and to preserve national room for maneuvering.

Since the end of the Cold War, conservative British leaders have reiterated their belief that trends in Europe have confirmed that their vision of a looser, broader, and more open EU will inevitably win out. London has, for example, sought to ally itself with Bonn to push for the broadening of the EU while allying itself with France on other key issues to balance German influence and to push the British vision of a looser Europe based on nation-states.[6] The opposition Labor Party under Tony Blair, on the other hand, has criticized the Tory government for isolating and marginalizing the United Kingdom in Europe and has pledged, if elected, to try to bridge the antagonisms that have existed between London and many continental EU members for over a decade now on these matters.

---

[6]Asked recently where the United Kingdom had demonstrated leadership in European affairs, Major responded that the most important development was the broadening of the EU, a development "made possible through an Alliance between Germany and Great Britain." (See "Wir bestimmen Europas Kurs," 1994.)

At the same time, London has also tried to maintain its role as the most privileged and predictable of Washington's NATO allies. It has found itself beset by weak government, unsure about its interest and willingness to engage in new trouble spots in the East and South, and often out of step with Washington on a number of issues—and thereby in danger of being eclipsed by Germany as Washington's key ally and partner in Europe. At a time when British and American policy seemed increasingly out of step on such issues as Bosnia, British diplomacy moved in the course of 1994 to realign itself with Washington on other fronts, such as NATO enlargement, to sustain a privileged relationship and to further London's own vision of an expanded Europe in a close relationship with the United States.[7]

THE EU'S UNCERTAIN FUTURE

These dramatic changes in Europe's strategic landscape have inevitably had a cascading impact on the key European and transatlantic institutions that have formed the core of the transatlantic relationship. The underlying assumptions and the implicit and explicit bargains on which these institutions depended were overturned by the course of events and their respective effects on the agendas and priorities of many key players. With the collapse of communism, much of the conventional wisdom on Europe's future that dominated the thinking in foreign and defense ministries, as well as the transatlantic strategic community, ended up in the trash bin of history.

Perhaps no institution has been more dramatically affected by these changes than the EU. In the immediate aftermath of the collapse of communism, the EU emerged as an institution seemingly destined to assume a role, if not the crucial role, in shaping Europe's future. With the continent's division overcome, Europe seemed on the verge of coming together. No institution seemed more destined to be the leader and motivator in knitting together a more civilized and peaceful Europe. In the eyes of its proponents, the EU would prove that a closer union could be achieved through an act of deliberate political will, not just through a shared fear of a threat. The benefits

---

[7]See, for example, the speech by British Defence Secretary Malcolm Rifkind (1995) in Brussels.

of EU-led integration, in turn, would trickle down to the new democracies in the East, which would eventually be able to become members of the club themselves.

The end of the Cold War, however, has paradoxically produced the breakdown of the Maastricht vision. That vision, which in conceptual terms predated the revolution of 1989, embraced the goal of monetary union by the end of the decade plus the advent of a common foreign and security policy—both of which were conceived as the final steps before political union (Steinberg, 1993). What initially appeared as a great leap forward, a qualitative jump in an integration process the likes of which had not been seen in some 40 years, was soon overtaken by real-world events on the ground.

Throughout its history, the European integration process has pursued three different yet overlapping agendas and visions. The first is that of reconciliation. This was the real impetus for European integration—the desire to overcome nationalism and old patterns of European conflict and to make war in the heart of Europe impossible, especially between France and Germany. The second agenda was one of economic prosperity—the realization that Europe had to integrate economically if it was to compete successfully and grow economically. The third vision was that of a Europe that would eventually gain control over its own political destiny and become a global force.

In a formal sense, therefore, the EU was not a creature of the Cold War. Its evolution, politics, and agenda were nevertheless shaped by the Cold War in major ways. Similarly, the end of the Cold War has again shifted the EU's evolution, politics, and agenda in ways that were inconceivable before 1989. Prior to 1989, the European integration process was widely seen in Western Europe as having succeeded in reconciling the western half of the continent with the past and having laid the basis for decades of economic growth and prosperity. In short, the first agenda has been successfully completed, and enormous progress has been made on the second agenda as well. The Maastricht Treaty symbolized the attempt to make a great leap forward in terms of deepening the quality of political and economic integration to strengthen Europe's ability to continue to manage this second agenda and to start laying the basis for the third:

Europe's emergence as a more unified political, economic, and—eventually—strategic actor.

Many commentators have termed Maastricht the last treaty of the Cold War precisely because it reflected an agenda premised on European integration in this old and divided Europe. Almost before the ink on the treaty was dry, the fall of the Berlin Wall forced the question of EU enlargement on the agenda; the divisive debate over the recognition of Bosnia-Herzegovina undercut the bold talk about a common foreign and security policy; and the viability of the treaty's centerpiece, European monetary union, was called into question by the crisis of the exchange rate mechanism which, at least in part, was tied to German deficit financing of unification.

No single event was more important in this shift than German unification. Politically, German unification upset the delicate political and psychological balance within the EU. Whereas the European Community had initially been founded at a time when the Federal Republic, France, Italy, and later the United Kingdom enjoyed roughly equal weight, a unified Germany with its own evolving agenda now emerged as the dominant political and economic power. Economically, the German need to pursue a forced-pace reconstruction of the new eastern states quickly skewed German fiscal and monetary policies, leading to a doubling of the German federal budget in five years and to higher German interest rates. The latter sent monetary shock waves throughout the continent, contributing to the European monetary system crisis in the fall of 1992 and the prolongation of the recession in Europe.

In short, by the time the Maastricht Treaty was ratified, many of the political and economic goals and timetables were being called into question. The unprecedented criticism of the Maastricht Treaty and elite and public opposition to it in such countries as France, Germany, and Spain, traditionally strong supporters of European integration, left many politicians in a state of near shock. The French endorsed the treaty by the most niggardly of margins; the Danes initially voted no, switching to yes only after obtaining major exemptions from the Maastricht provisions. The debate in the United Kingdom only confirmed London's deep opposition to the tightly integrated union favored in Brussels. In Germany, the government was fortunate to have a constitution that did not allow a public referendum. Although the Bundestag voted overwhelmingly in favor of the treaty, it also added a final provision that it would vote again, if and when the day came, on the "finality" of European monetary union—in short a yes without consequence where it really mattered.

The second pillar of Maastricht—the creation of a common foreign and security policy—appears to have been badly wounded, perhaps mortally, in the Persian Gulf and Yugoslav crises. The Gulf War underscored Europe's foreign policy impotence and shattered the belief that the post–Cold War world would be one in which economic power would inevitably rule supreme over military might. During the Gulf War, Belgian Foreign Minister Marc Eysken's comment that Europe had shown itself to be an economic giant, a political dwarf, and a military worm encapsulated just how long the path to European defense identity remained.

The nightmare of genocide in Bosnia, coupled with the vastly different reactions to the crises within Europe, showed that nationalism in Europe was not dead and reawakened fears about a possible renationalization of European politics. Although the crises in the former Yugoslavia were initially proclaimed as the "hour of Europe," the lack of political will, internal divisions, and foreign policy impotence of the EU were soon demonstrated. Policy failures in one realm only reinforced difficulties in others. As Wolfgang Schäuble (1995, p. 72) put it in a roundtable discussion in March 1995:

> How am I supposed to convince my people that they should give up the D-Mark [for the cause of European integration] when Europe is incapable of stopping a war taking place at our doorstep? We shouldn't fool ourselves. Europe's inability to pursue a common policy in the former Yugoslavia, let alone to stop the war, is the disaster of European policy of the 1990s.

By the mid-1990s, it was clear that the EU was being driven in new directions under the impact of the geopolitical reshuffle taking place on the continent. Although Maastricht's architects believed that they could complete their designs for deepened integration within the EU prior to addressing the broadening issue, the combination of unexpected obstacles toward deepening plus growing pressures to open the EU to the North and East meant that the window for completing the Maastricht vision was rapidly closing. By early 1995, the EU had broadened to include Austria, Finland, and Sweden. At the EU summit in Essen in December 1994, the EU agreed to a preliminary strategy for extending membership to a total of ten states in Central and Eastern Europe.

Whereas broadening was becoming a political reality, prospects for deepening remained clouded with uncertainty. Within a few short years, the EU had moved from a vision of deepening integration between a fairly closely knit 12 members in Western Europe to the prospect of a pan-European EU with some 25 to 30 members. Under the pressure of events, the EU was being forced to return to an agenda of reconciliation, this time the reconciliation between western and eastern Europe and the creation of a new Europe. A unified Germany was in many ways the driving force behind this trend. Traditionally both the financier of and a key political broker in the European integration process, Germany's policymakers quickly concluded that they had a powerful set of moral, political, economic, and strategic motives for turning the EU into a pan-European institution. At the same time, German leaders continued to argue that enlargement could not take place at the cost of deepening, with Helmut Kohl going so far as to term European integration a question of "war and peace" for Europe's future.

The prospects for such integration and for the EU emerging as a more cohesive actor were clearly deteriorating. To be sure, many commentators view the EU's current problems as temporary, largely a function of Europe's current economic recession, the unexpected costs of German reunification, overly ambitious plans, and a loss of political confidence. A growing number of politicians and commentators, however, are challenging the very logic that has driven the European integration process since the 1950s. As Sir Ralf Dahrendorf, currently Warden of St. Antony's College at Oxford University and previously an active spokesman and supporter of European integration as both a theoretician and politician, recently wrote:

> The European unification process has come to a standstill. Maastricht marks the end of a process, not a new beginning. It has divided the citizens of the European Community into pro and anti-Europeans, their governments into those who favored an accelerated approach and those who wanted to take their time. The phrase

European Union comes at an inopportune time, for Europe is less unified than at any time since the late 1950s.

> Maastricht was not only a treaty that divided, it also seemed like a gigantic matter of secondary importance in light of the new problems confronting Europe: the opening of the former communist states of the East requires a decisive and positive answer in the West; the ongoing recession is creating problems of competitiveness, employment and social cohesion which all the countries of Europe must address; and the changed international political geography demands decisions by Europe. In light of these challenges it is almost silly to continue along the same old path—and hardly surprising that the citizens of Europe perceive it as such. (Dahrendorf, 1994, pp. 18–19.)

While one can hardly imagine that any of its current members would actually abandon it, momentum for European integration has clearly eroded. The emergence of a tightly integrated and coherent EU emerging as a cohesive political and strategic actor capable of managing its own security in Europe, as well as acting as a global partner of the United States has, like a political mirage, again receded into the distance.

The key test will come, of course, in 1996, when the EU members meet for the next IGC to determine the future path for deepening integration, above all economic and monetary union, as well as future prospects for a common foreign and security policy. Maastricht II looms increasingly large as a watershed as the EU struggles to come to terms with the needed reforms of decisionmaking mechanisms within the EU, as well as the geopolitical consequences of eastward expansion and growing pressures for a more active role in the South.

The prospects for success, however, are far from certain. The original hope of achieving Maastricht's goal of monetary union by January 1997 has de facto been dropped. Even the fallback position of achieving monetary union by January 1999 is likely to be possible only for a small handful of countries. National preferences and visions for the EU's future are more different today across Europe than they have been in years. Even between such close partners as France and Germany, there are substantial differences over whether and how to deepen integration. Variable geometry, multitrack, multispeed, two-tier, hard core, concentric circles, à la carte—this list of metaphors reflects just how divergent the views of various EU members are. The possibility of gridlock or political paralysis should not be underestimated. As Stephen Graubard (1994, p. xi) wrote in the introduction to an issue of *Daedalus* devoted to the question of the future of Western Europe—and appropriately entitled "Europe Through a Glass Darkly"—there is a new uncertainty among EU countries about what steps now need to be taken to achieve greater unity, or indeed whether such unity is as important as was once imagined. The EU's uncertain future is a mirror image of the growing uncertainties facing Europe as a whole.

QUO VADIS THE ATLANTIC ALLIANCE?

The impact of the end of the Cold War on the Atlantic Alliance has been even more dramatic. The collapse of communism, the unraveling of the former USSR, and the withdrawal of the troops of the former USSR some 1,000 km eastward removed any direct military threat to Alliance territory. It also seemed to remove the most direct and immediate rationale for U.S. military presence and, indeed, for the very existence of NATO—the deterrence of a direct Soviet military threat to Western European territory.

Both the United States and its European allies in the Atlantic Alliance were suddenly confronted with questions that had been fiercely debated for years by academics in theoretical terms and carefully avoided by politicians: Was the U.S. presence in Europe destined to be a temporary protectorate or a permanent feature of the European landscape? What would happen if the Americans went home? Would Europe coalesce in new ways or fragment? In the 1980s, Joe Joffe termed the U.S. military presence on the continent "Europe's pacifier." So long as the U.S. military presence in Europe persisted, he argued, the Europeans would succumb to the comfort of the pacifier and never undertake a serious effort at European defense. Only when that pacifier is removed might Europe start to grow up in security and defense terms and be able to handle its own security.

Had the time come to test this theory? Both Washington and its key European allies quickly concluded that the answer was no. Washington feared the prospect of being shut out of Europe in both security and economic terms. Faced with the unsettled future of

Russia, their own doubts about their ability to manage European security on their own, a unified Germany, fears of renationalization, and the new uncertainties of the post–Cold War landscape, Washington's European allies also quickly opted for the same answer.

In the immediate aftermath of the end of the Cold War, therefore, the debate over the Alliance's future focused on how to sustain NATO after the collapse of the Warsaw Pact and, subsequently, the USSR. The Alliance moved to embrace a new concept for NATO and the U.S. military presence that transcended the Soviet threat and could thus survive its disappearance. It embraced the rationale of the United States as a European power with the same commitment, presence, and voice in European security affairs as the Europeans themselves. The United States was no longer Europe's protector but rather a permanent participant in European affairs.

Consequently, the role of the Atlantic Alliance was broadened and redefined. NATO was to expand its responsibility from that of defending Western Europe to that of managing security in Europe as a whole. Both rationales were substantially different from those initially advanced by the founders of the Alliance in the late 1940s. Alliance communiqués from 1989 to the present document the step-by-step redefinition of the Alliance's political rationale and its conceptual underpinnings. To be sure, even during the Cold War, the Alliance had de facto assumed several secondary missions in addition to deterring a direct Soviet military threat. Such tasks as promoting transatlantic solidarity, avoiding renationalization, and promoting integration or ending the division of Europe had long come to be seen as key benefits flowing from the very presence of a strong Atlantic Alliance.

However, these tasks were now moved into the forefront of NATO's purpose and official rationale. In addition, NATO adopted a whole series of new missions, including peacekeeping, the projection of stability to the East, and a proposed new dialog engaging non-NATO countries along the Mediterranean littoral. Also on the list was crisis management—a term deliberately left ill-defined as a code for NATO's possible future involvement in conflicts beyond NATO's borders. NATO communiqués started talking not only about the inviolability of the borders of Alliance members but also of the "inseparability" of security of all states in Europe. NATO had already, in Rome in November 1991, anointed itself as the guardian of European stability and officially accepted a new strategic concept reorienting its military forces away from border defense toward rapid reaction and projection. In June 1992, peacekeeping was officially added to the list of new NATO missions. This was followed, at the January 1994 summit, by the decision in principle to enlarge to the East, adopt an outreach initiative in the Mediterranean and, finally, an Alliance initiative on proliferation.

To be sure, the Alliance also went out of its way to make room for a strengthened European pillar as well as for the possibility that the Europeans might act independently from the United States. The Alliance's proposal for Combined Joint Task Forces and forces that were "separable but not separate" was carefully crafted to make such moves possible. The Alliance was also careful to ensure that participation in new non–Article 5 missions was voluntary, thereby creating for itself and other countries the option of opting out if and when it was decided that vital U.S. interests were not involved. The entire package was then subsumed under the concept of interlocking institutions that could provide both flexibility and effective security in the new Europe.

While elegantly resolved in paper communiqués and in the minds of U.S. and European policy planners, the true ability of the Alliance to modernize itself has been tested in its response to real-world challenges. The debate over NATO enlargement and the broader issue of NATO's role beyond Alliance borders, which was catalyzed by the Bosnia crisis, underscores how difficult it has been to implement these changes in practice.

The debate over NATO enlargement was publicly opened in the spring of 1993 by German Defense Minister Volker Rühe. Speaking at the IISS's Alastair Buchan Memorial Lecture—the same forum in which Helmut Schmidt had launched the INF debate in 1977 that would dominate the Alliance for the next seven years—Rühe argued that NATO's basic problem was the mismatch between its old mission and the new strategic challenges in and around Europe. It was, therefore, no longer tenable for NATO to concentrate on the "strategic luxury" of territorial defense of its current members. The distinction between so-called "in area" and "out of area" crises was artificial and anachronistic in the new Europe after the Cold War. The Alliance, the German Defense Minister noted, must not be a "closed shop" but rather must be open to new members. Above all, it must project stability eastward to protect the fragile democracies of East-Central Europe and guarantee their success. Eastern Europe cannot remain a "strategic no-man's land" fraught with potential conflict and instability. Membership in NATO for these countries, in Rühe's words, could come even prior to their joining the EU.

In the summer of 1993, this debate moved from an insiders' discussion into the public domain. In June 1993, U.S. Senator Richard Lugar, in a speech before the Overseas Writers Club, called for a much broader redefinition of the Atlantic Alliance and a new strategic bargain between the United States and Europe to deal with the emerging new threats in Europe. The Alliance, in his words, had to go "out of area or out of business." (See Lugar, 1993. See also Asmus, Kugler, and Larrabee, 1993.) In the run-up to the January 1994 NATO summit, a brief but intense debate ensued in the American and transatlantic political arenas over the future U.S. role in European security, the future functions of the Atlantic Alliance, and Western policy toward Russia.

The administration's "Partnership for Peace" (PfP) proposal was born as a simultaneous attempt to address legitimate security concerns in East-Central Europe and not to isolate Russia in a way that could undercut the reform process and lead to an anti-Western backlash. At the same time, the administration committed itself to NATO expansion in principle. Speaking at the North Atlantic Council summit, President Clinton (1994a) stated that PfP was designed "to set into motion a process that leads to the enlargement of NATO." Speaking at a press conference in Prague two days later, Clinton (1994b) affirmed that PfP "changes the entire NATO dialogue so that now the question is no longer whether NATO will take on new members, but when and how."

While the PfP compromise bought the Alliance time, the real issue was when the Alliance would move beyond the question of "whether" and start to address the questions of "when and how" regarding expansion. From the outset, it was clear that these questions were linked. Any attempt to build a consensus for expansion either in the United States or across the Atlantic was contingent upon the details of how expansion might take place and what costs were involved. In the summer and fall of 1994, Washington shifted policy gears, and U.S. officials signaled their desire to start intra-Alliance negotiations on the enlargement process, leading to the decision at the December 1994 Ministerial to initiate a study within the Alliance, to be completed by fall 1995, on the when and how of eventual expansion.

The issue of Russia and its role in future European security also continued to loom large in the internal NATO debate. A consensus had crystallized in the West that Russia was not eligible for either EU or NATO membership. Not only was Russia simply too big and different, but the broader question of Europe's geopolitical balance was at stake. At the same time, nearly all Western countries continued to bend over backward not to take any steps that would further isolate the Yeltsin government, for fear that such steps might play into the hands of his more conservative and nationalistic political opponents. As the consensus in Russia against NATO enlargement solidified, however, NATO increasingly faced a conundrum. Although the Alliance repeatedly underscored that enlargement was not aimed at Moscow, every small step taken in the direction of enlargement seemed to produce even harsher rhetoric from Moscow—with Yeltsin warning in the fall of 1995 that enlargement could "light the fires of war all over Europe."[8]

This increasingly raised the question whether the Alliance and Russia were on a collision course over this set of issues. Officially, Washington and other key Western capitals insisted that Western policy could strike a balance between NATO expansion and offering Moscow a "privileged partnership" that would address Moscow's concerns. At the same time, Moscow is increasingly adamant in its rejection of the premises or the logic of Alliance policy on NATO enlargement and appears increasingly prepared to raise the ante in its efforts to undermine or stop the enlargement process. The Alliance's initial assumption that going slow on enlargement would give the Russians time to get used to it may have been false. As the consensus against enlargement hardens in Moscow, a policy of

---

[8] *New York Times*, September 11, 1995.

"going slow" may simply offer Russian diplomacy more temptations and opportunities to try to undercut the process.

Moreover, the more Moscow attempts to stop enlargement, for example, the harder it becomes to build a consensus within the Alliance for a meaningful cooperation package as old fears about Russia's intentions are reinforced and concern increases about potentially opening the door to Russian influence at a time when Russian policy is moving in the wrong direction. Ultimately, Alliance leaders have to decide just how far they are willing to go in attempting to persuade Moscow that the West is not trying to isolate Russia—irrespective of the fact that Russian policy on nearly all aspects of NATO's transformation is increasingly hostile. If NATO remains timid in its approach, the chances of Western policy having any substantial impact on Russian attitudes is low. At the same time, trends within Russia itself and in Russian policy toward the West make it increasingly difficult to put together a meaningful cooperation package lest such steps divide the Alliance and run the risk of appearing as if the West is only making "concessions" in the face of Russian coercive pressure. Ultimately the Alliance may have to choose between its commitment to enlargement and relations with Russia.

In the public eye, the NATO enlargement debate has increasingly been overshadowed by the Alliance's role in the war in Bosnia—the first actual involvement of NATO in combat operations since its founding in 1949. Although the Bosnian crisis and the West's involvement are beyond the scope of this chapter, the ramifications of Bosnia for the future of both Europe and the Alliance are essential. For many Americans, it has raised the issue of whether such conflicts really threatened vital U.S. interests and just how engaged the United States should become in the problems and conflicts of the new Europe. To be sure, the United States had no legal obligation to intervene; it simply made use of its opt-out option carefully codified under NATO's new strategy. Anyone watching the U.S. congressional debate could hardly escape the confused deliberations over just when and where U.S. vital interests were at stake. For many Europeans, in turn, the debate raised the question of just how reliable a partner the United States would be in the future. The question could hardly have come at a more inopportune time, because most Europeans had de facto concluded that building an independent European defense without the United States was neither desirable nor feasible.

The result was a growing paradox. On the one hand, hardly a week passes without senior officials in the United States and Europe lauding the importance of the transatlantic connection, emphasizing the indispensability of the Atlantic Alliance and committing themselves to modernizing the transatlantic bargain to meet the requirements of the post–Cold War era. Today in Europe there are no signs of anti-Americanism and no political forces of any significance questioning the U.S. presence. Even Paris, Washington's old nemesis in the struggle for leadership in Europe, has quietly shifted gears to bring itself closer to the Alliance and to expand cooperation with Washington to ensure that the United States remains engaged in European affairs. Pro-American sentiments may even be higher in Eastern than in Western Europe. All of Europe, it seems, has truly accepted the United States as a permanent European power, not just a temporary protector during the Cold War.

On the other hand, the oft-repeated official affirmations about the ongoing necessity of the transatlantic relationship also reflect the unofficial but nevertheless very real uncertainty about its long-term viability. If one looks beyond the facade of unity presented by transatlantic officialdom, the reality is that there are growing doubts on both sides of the Atlantic about the future of this relationship— albeit for different reasons. The Bosnian crisis has revealed a substantial divide between U.S. and European views and priorities and has led to gnawing doubts about whether Alliance members are reliable. In the eyes of many Europeans, the real question in the future of European security is less Russia's uncertain future than whether America is willing and able to play the role the Europeans clearly want it to play.

In the United States, a growing chorus of voices questions how relevant this relationship is to current U.S. interests and priorities. These voices include not only the siren calls of isolationists of the Right and the Left, but also a growing number of mainstream figures from the bedrock of traditional pro-Atlanticist support, who now wonder out loud whether the transatlantic relationship in its current form is sustainable or even desirable in light of the dramatic changes that have taken place in Europe and shifting U.S. interests and priorities globally. No longer consumed by Soviet military power, Americans are increasingly worried about non-European threats. Europe's insularity and strategic myopia raise questions about its effectiveness as an ally and partner and lead some to conclude that the United States should simply go it alone.

Jim Thomson of RAND (1995) posed the right question when he proposed a *Gedankenexperiment* imagining two worlds, one in which NATO exists and one in which it does not. If NATO did not exist today, would we want to recreate it? If so, in what form and for what purpose? There clearly are some powerful reasons why the United States should want a close transatlantic relationship. Having fought two world wars this century and spent nearly one trillion dollars in ensuring European security during the Cold War, the last thing the United States can afford today is an unstable Europe. In addition, more than ever before, the United States needs coalition partners to play an effective role in global affairs and international security. By building effective coalitions with like-minded countries, the United States can maximize Western influence and save resources. Europe is our natural partner.

Listing these factors, however, already underscores how the American agenda in the Alliance is shifting. While the United States obviously wants to ensure a stable Europe, Europe no longer dominates the post–Cold War strategic calculus in the same way as it did during the Cold War. This has less to do with U.S. isolationism than the simple fact that Europe's importance to the United States increasingly depends on its ability to play the role of an effective partner in addressing the post–Cold War strategic agenda. As Thomson has pointed out, if the United States had to renegotiate the transatlantic security bargain today, it would certainly still want a close relationship but it would have little interest in negotiating a replica of today's relationship. This core question—what transatlantic relationship and Alliance in what form and for what purpose?—is one that both sides of the Atlantic have to answer not only in terms of rhetoric but in reality.

CONCLUSION

The collapse of communism and the unraveling of the former USSR have had a cascading effect throughout the politics of Europe. The aftershocks of this geopolitical earthquake are likely to be felt for some time to come. While the impact is already visible in some countries, these changes are only starting to work themselves through the political systems of others. What is striking—and worrying—about Western Europe since the end of the Cold War, however, is not how the initial euphoria and optimism regarding the future have dissipated, for that was in many ways inevitable. Rather, it is the loss of self-confidence.

The sense of initiative and vision that came to the fore in Western Europe in the initial aftermath of the collapse of communism has dissipated when it sought to confront problems ranging from unemployment, Russia, and integrating Eastern Europe to war in the former Yugoslavia. Even in the more prosperous Western European countries, there is a growing sense of doubt in the ability of governments to ensure the domestic prosperity and social protection on which Europeans have prided themselves, as well as doubts even in the ability of the West to prevent racism, war, and genocide from returning to the continent. Things that many Europeans assumed to be part of their past have now returned as reality—with governments either unwilling or unable to prevent them. The optimism that experts could be engaged and institutions could be built that would address and fix these problems has faded.[9]

To be sure, some would argue that the situation is far less dramatic, that Western Europe has only temporarily lost its bearings, that it remains essentially on track, and that, its current problems notwithstanding, it will continue to muddle through in the years ahead. However, it is also noteworthy that small but growing numbers of commentators are increasingly asking whether the crisis facing Europe is more than a Balkan crisis, or even a crisis of fragile democracies in Eastern Europe or Russia, but rather whether Europe might

---

[9]As Stephen Graubard (1994, p. xv), the editor of *Daedalus*, concluded in the above-mentioned issue of that journal:

> It is as if the whole of the Atlantic world, and not only the former Marxist territories, are hopelessly bewildered. The easy explanation for the condition is that the problems of these societies have never been greater. The more terrifying possibility is that they are not being correctly perceived, that slogans too frequently substitute for policy, and that neither governments nor their oppositions offer viable solutions for the muddles that are seen to exist.

not be slipping into something akin to the 1920s and 1930s—plagued again by indecision and the lack of a moral compass preventing it from effectively confronting internal weaknesses or foreign policy aggression (see Sommer, 1993).

The answers to these questions clearly have far-reaching consequences for U.S. national security strategy. Growing instability in and around Europe could again turn Europe into a major national security concern for the United States in the years ahead. As Washington watches Europe's new strategic landscape develop, how should it define and prioritize U.S. interests? The old status quo is simply dissolving. The United States increasingly faces the choice of reaffirming its security role in a very different strategic context or becoming irrelevant to Europe's new problems.

Although many commentators may insist that these new challenges are not crucial to U.S. interests and should be left to the Europeans, the reality is that the hope—or fear—of Europe developing the coherence and capability of managing its security on its own has come and gone. Europeans—including the French—know that Europe today is not capable of addressing the new strategic challenges in the East and South without active and strong American leadership. Indeed, Washington's own policies will undoubtedly be one factor determining which way Europe goes in the years ahead.

The prospect of renewed instability in and around Europe means that it may reemerge as a theater for U.S. defense planning in ways not originally envisioned after the end of the Cold War. Europe did not, for example, play a major role in recent U.S. defense planning exercises, such as the Bottom Up Review. In the years ahead, it may again become a theater for U.S. defense, not necessarily because major conflicts or war loom, but simply because a new geostrategic drama is unfolding that will also be shaped and influenced by defense preparations. Whether and how the United States engages in this new Europe, however, will also be tied to whether the United States and Western Europe can reach a new set of political and military understandings that also address the common challenges and possible threats they face beyond Europe.

As Europe's new geostrategic drama plays itself out in the years ahead, the United States inevitably remains a key actor in it. For the

United States, the key issue is whether future instability in and around Europe's periphery is seen as central or peripheral to American national security interests. Should instability and disorder in Europe increase in the years ahead, the United States may again be confronted with a divisive debate over whether and how to define its future engagement in an increasingly unstable Europe. The Cold War may be over, but the challenge of creating a stable political order in Europe is not.

REFERENCES

Asmus, Ronald D., *German Strategy and Public Opinion After the Wall 1990–1993*, Santa Monica, Calif.: RAND, MR-444-OSD/AF/A, 1994.

_____, *Germany's Contribution to Peacekeeping: Issues and Outlook*, Santa Monica, Calif.: RAND, MR-602-OSD, 1995.

Asmus, Ronald D., Richard Kugler, and F. Stephen Larrabee, "Building a New NATO," *Foreign Affairs*, September–October, 1993, pp. 28–40.

Balladur, Édouard, "For a New Élysée Treaty," *Le Monde*, November 30, 1994.

Clinton, W. J., "Remarks by the President at the Intervention for the North Atlantic Council Summit in Brussels," Washington, D.C.: Office of the Press Secretary of the White House, January 10, 1994a.

_____, "Remarks by the President at the Press Conference with Visegrad leaders at the U.S. Ambassador's Residence in Prague," Washington, D.C.: Office of the Press Secretary of the White House, January 12, 1994b.

_____, "President Clinton's Speech to the French National Assembly," *Foreign Policy Bulletin*, Vol. 5, No. 1, July/August 1994c.

Dahrendorf, Sir Ralf, "Ein Europa für die Zukunft," *Der Spiegel*, No. 1, 1994, pp. 18–19.

Freedman, Lawrence, "Strategic Studies and the New Europe," *European Security after the Cold War*, Adelphi Paper No. 284, 1994, pp. 15–27.

Graubard, Stephen, preface to "Europe Through a Glass Darkly," *Daedalus*, Spring 1994.

Hamilton, Daniel S., *Beyond Bonn: America & the Berlin Republic*, Washington, D.C.: Carnegie Endowment for International Peace, Carnegie Endowment Study Group on Germany, 1994.

Hoffman, Stanley, "Europe's Identity Crisis Revisited," *Daedalus*, Spring 1994, pp. 1–25.

Joffe, Josef, "The New Europe," *Foreign Affairs*, No. 1, 1993. p. 43.

Juppé, Alain, author's personal copy of a speech delivered on the 20th anniversary of CAP, at the Centre d'Analyse et de Prévision (CAP), Paris, January 30, 1995.

Kohl, Helmut, *Bulletin*, No. 58, Press and Information Office of the Federal Government, Bonn, July 14, 1995.

Larrabee, F. Stephen, *East European Security After the Cold War*, Santa Monica, Calif.: RAND, MR-254-USDP, 1994.

Lellouche, Pierre, "France in Search of Security," *Foreign Affairs*, Spring 1993, pp. 122–131.

Lugar, Senator Richard, "NATO: Out-of-Area or Out of Business," speech delivered to the Overseas Writers Club, Washington, D.C., June 24, 1993.

Rifkind, Malcolm, British Defence Secretary, speech delivered to the Belgian Royal Institute of International Relations, Brussels, January 30, 1995.

Rühe, Volker, *Deutschlands Verantwortung: Perspektiven für das neue Europa*, Frankfurt-am-Main: Ullstein Verlag, 1994.

Sommer, Theo, "Die Krise holt den Westen ein," *Die Zeit*, No. 15, April 9, 1993.

Schäuble, Wolfgang, quoted in *Die Verfassung Europas, Bergedörfer Gesprächskreis 103*, Hamburg-Bergdorf: Körber Stiftung, 1995, p. 72.

Steinberg, James, *An Ever Closer Union: European Integration and Its Implications for the Future of U.S.-European Relations*, Santa Monica, Calif.: RAND, R-4177-A, 1993.

Thatcher, Margaret, *The Downing Street Years*, New York, N.Y.: Harper Collins Publishers, Inc., 1993.

Thomson, James A., "Back to Square One with Einstein," *Los Angeles Times*, March 21, 1995.

"Wir bestimmen Europas Kurs," interview with John Major, *Der Spiegel*, No. 17, April 25, 1994, pp. 24–28.

Chapter Three

RUSSIA
Fritz Ermarth[1]

INTRODUCTION AND BACKGROUND

A strategic assessment of Russia is essentially an assessment of its ongoing revolution. How this revolution proceeds and turns out represents the most profound security issue posed by Russia for the United States and its allies.

Russia's revolution is as sweeping as that which occurred at the beginning of this century, although quite different from it. Revolutions usually involve the takeover of a country's political institutions by a new elite or class, who then may remake the institutions governing society and thus effect broader social changes. Russia's current revolution has put everything on the block of change at once, except, ironically, its ruling class. The vast majority of the politicians now seeking to remake Russia were card-carrying members of the old communist nomenklatura who have remade themselves according to various guises and convictions ranging from democratic to fascist. Under them, Russia is seeking to construct wholly new political institutions at the nation's center and at all levels of regional and local power. It is attempting to build an economy on market principles amidst the institutional and behavioral detritus

---

[1]Fritz W. Ermarth is currently a senior officer at the Central Intelligence Agency. This chapter is based on research and writing he did while a sabbatical fellow at RAND in 1993–1994, and on subsequent updates and editings. None of its contents should be attributed to CIA or the U.S. Government.

of a sclerotic administered economy. It is trying to build a civil society after centuries in which society independent of the state was always weak and after 70 years in which it did not exist. It is forced to redefine the relationship of Russia and Russians with the rest of the world, with other peoples within its borders, with neighboring former colonies of a continental empire, and with the world at large. The historical legacies of autocracy and violence that inform this experiment are forbiddingly negative. Even the physical setting of broken-down infrastructure, obsolescent capital stock, pollution, and calamitous public health is daunting.

As of late 1995, the Russian revolution has a mixed record, and its prospects are very uncertain:

- It has been the Western and Russian hope that the revolution would result in a Russia that is a normal—nonimperial—and democratic country. For the first several years of the revolution, the indicators were mostly positive. There was very little political (as opposed to criminal) violence in central Russian politics. Then the October 1993 clash between Yeltsin and the legislature occurred. It was hoped that this violence would be an exception. Then Chechnya took place. Russian brutalities there dampened optimism about Russia's prospects. Uncertainties about whether the presidential and parliamentary elections would take place—and should they take place with what results—have clouded politics throughout this period.

- Russia remains very unstable. Some key federal institutions are weakening. The role of criminal elements in society and even state institutions has grown. Until 1993, foreign and Russian observers thought the breakup of the country was a serious, indeed likely, possibility. In 1994, this threat receded. In 1995, the war in Chechnya raised fears that this threat had become stronger. Important questions about the safety and security of nuclear weapons, nuclear materials, and other critical technologies in Russia remain.

- Following parliamentary elections in December 1993 in which communists, extreme nationalists, and authoritarians did unexpectedly well, Yeltsin formed a government run by former communist managers whose common platform was railing against the evils of rapid economic reform. But the predicted retreat from reform and inflationary spending did not occur: Spending and inflation were kept down; privatization continued; bankruptcies began; and overall reform was still on course. But in the aftermath of the backlash elections of 1995 and going into the presidential campaign of 1996, the future of economic reform is once again in doubt

Russia is still struggling to find a place for itself in the world. After a period of emphasizing cooperation and partnership with the United States and the West, Russian foreign policy has become more nationalist and assertive and less cooperative with the West. There has been a trend toward disillusionment with the United States. From implementing strategic arms agreements, to the sale of reactors to Iran, to disagreements about sanctions on Iraq, Bosnia, and NATO expansion, Russian-Western relations are becoming more contentious.

The Russian military continues to weaken—although Russia remains Europe's most heavily armed power. The role of nuclear weapons in Russian military doctrine has grown. The performance of the Russian military in Chechnya has weakened its standing in Russian society and its claim to resources. The Russian economy has declined, and its prospects for recovery are uncertain, although it appears to be headed for an upturn in 1996.

The old communist order has been irreversibly destroyed. Russian society, politics, and economics are still undergoing processes of breakdown or deconstruction. Even the rosiest optimist must recognize that the path to a stable result of any kind, and especially a democratic result, is highly uncertain given the existence of formidable obstacles. The processes dominating the scene paint a very mixed picture.

TIME DIMENSIONS

It cannot be stressed enough that the current Russian revolution is only in the beginning stages of what is certainly going to be a long travail. The first Russian revolution of this century had its roots in the 19th century. It took until 1939, or some would argue even 1948, to create the final result, full-blown Stalinism.

The second Russian revolution of this century had its roots in the nature of Stalinism itself. Stalin created an unreformable political order; it could not renew itself, except, as some early students of the system, like Brzezinski, observed, through periodic purges of the ruling class, which that ruling class preferred not to undergo. It also rejected the bloodless substitute of bureaucratic and cadre turbulence that Khrushchev tried to introduce. It is hard to date when real dissidence began in the Soviet system, because the system had a few visible and many secret opponents throughout its history. That the system was fundamentally defective began to intrude upon the consciousness of Soviet intellectuals as early as the mid-1950s. The propensity of the Soviet economy toward stagnation was visible by the end of the Khrushchev period. By the end of the 1960s, political dissidence had emerged in earnest, and Western analysts of the Soviet order could seriously contemplate the alternatives between evolution and collapse. In the 1970s, effective suppression of the dissidents, windfall oil profits for the economy, détente and Western acceptance, and the growth of strategic military power seemed to revive prospects for the Soviet system. But by the late 1970s and early 1980s, its stagnation became ever more visible and dramatic.

There is debate about how much the efforts of the West to contain the power of Soviet communism, and particularly the efforts of the Reagan administration to challenge it in the 1980s, contributed to the demise of the Soviet order. On an international scale, the Soviet system seemed on the march in the late 1970s, but blocked and decaying by the mid-1980s. International and domestic realities played together in the complex chemistry of the breakdown of the old Soviet regime. The Gorbachev phase has yet to be fully captured by historical analysis. He attempted to reform the system, but, through a combination of boldness and naiveté, he destroyed it. The timing may have been accidental. Had the disciplinarian Andropov lived longer, so probably would have the Soviet Union and its Communist Party. But the result we witnessed between 1985 and 1991 was sure to come, although it might have been postponed for one or more decades.

So, like its predecessor, the current Russian revolution was long in coming and will most assuredly be long in playing out. Hence, Yergin (1995), in his very insightful *Russia 2010*, has not picked too long, but perhaps too short, a time horizon. Strategic planners should note that, after the Bolshevik revolution, it took 30 years, a world war, and the destruction of Germany and Japan for Soviet Russia to emerge as a serious strategic threat to the West. Whether the quickening pace of change augurs for a quicker result in the case of the Russians in our time remains to be seen.

CONFUSION

It is important to appreciate the difficulties of comprehending the current Russian scene. Russia is more open to scrutiny than perhaps ever before in its history. Yet it remains difficult to derive an accurate, comprehensive, and balanced view of its affairs.

Secrecy is relaxed greatly, but not absent. To test this point, one might seek, for example, the sources of party campaign funds or the true casualties of the October 1993 events or the current status of Russian chemical and bacteriological weapon programs.

The media of Russia are freer than ever, as shown by the coverage of the ongoing war in Chechnya, but not totally free. Many organs of the media are still imbedded in official institutions. Many more are still dependent on official subsidies, lacking an adequate base of self-financing through subscription and advertising. Until Russia develops a powerful free market and the rule of law, the freedom of the media will be at risk.

Official economic data are of dubious value. Never completely reliable in the Soviet period, they are even less so now, but in the opposite direction. Soviet officials had every interest in overreporting their performance. Now Russians of all stripes have powerful incentives to underreport performance, to avoid payment of taxes to the government or of subventions to organized crime. In an economy and society that are rapidly experiencing "unofficial" development, this is a powerful source of distortion. At present, the basic problem is that private economic activity—rapidly growing and surely to be counted as one of the basic factors of Russian development—is impossible to measure and very hard to estimate.

Polling of public opinion or of the views of various subgroups and elites is increasingly noted on the Russian political scene. The analytic professionalism of Russian pollsters is increasing. But it is hard to create valid samples in Russia when population data are uncertain. It is hard to get valid interviews when the phone system is unreliable and when interviewees may still be afraid of expressing their opinions openly. Recently, some Russian officials have expressed their fear that the security agencies might have turned on the tape recorders once again, and some officials prefer not to talk to their foreign friends in their offices.

Clausewitz spoke of the fog of war. Those who seek to understand what is going on today in Russia must peer through the fog of revolution. To put the matter in more prosaic terms, Russia has yet to develop the integrated and reliable information infrastructure that would enable domestic and foreign observers to gauge its affairs with some sense of confidence in the comprehensiveness and accuracy of information. A stable and reliable information infrastructure in a society is a sign that the society is stable and reliable. We may not confidently understand what is going on in the Russian revolution until it is truly over—which might take many years.

POLITICS

In the year following Yeltsin's dissolution of the old parliament and the armed clash that ensued, there were a number of, on the whole, positive developments in Russian politics. This surprised both Russians and foreign observers because the outlook was rather grim in the aftermath of the October 1993 events.

The elections of December 1993 were on balance a positive development despite low voter participation and the poor showing of reformers. They were fair and produced a more representative and legitimate legislature than the previous one. Among members of the new Duma, opponents of reform are numerous. But they, like the reformers, were divided ideologically and fractious personally. Under a moderate and relatively skillful leadership during its first year, the Duma was more effective than its predecessor. A budget and some dozen laws were passed and signed by the president, and deadlock between legislature and executive, as witnessed in 1993, was avoided.

Yeltsin himself did somewhat better than expected in 1994. He generally eschewed confrontational tactics. Expectations that poor health might bring a premature end to his term receded. He seemed in charge and appeared more likely to remain so than he did at the turn of the year.

Prime Minister Chernomyrdin ran a moderate, effective, and unexpectedly proreform government. It relied heavily on the president's great powers to govern by decree, but managed where necessary, as on the budget, to get cooperation from the Duma.

Russia's adoption of a real constitution through referendum in December 1993 and its basic acceptance by the political spectrum was an important development. The constitution asserted civil, human, and political rights. It provided the legal basis for private economic activity and property, including land ownership. It ordained an independent judiciary (with much implementation yet to be accomplished). It provided for a strong federal structure, but in principle granted more power to the regions than ever before in modern Russian history. It also granted strong powers to the president. He can issue decrees with the force of law. The Duma can override them by passing contrary laws by a two-thirds majority. He can appoint ministers without the Duma's consent, but the Duma must approve the prime minister. He can dissolve the Duma and call new elections under specific conditions of no confidence appropriate to a parliamentary system, but is prohibited from doing so under others when aggrandizing personal power might be his aim. He cannot dissolve the upper house, the Federation Council, under any circumstances. Without doubt this constitution could be abused by an autocratically inclined president.

A second major event, which Yeltsin intended to be of nearly constitutional importance, was the agreement on Civic Accord in April 1994. Not a law, but rather a pledge or mutual exhortation of its adherents, the agreement was expected to have real political and psychological significance. In essence, it was a promise by the signatories to avoid the confrontational tactics that produced deadlock and then violence in 1993. A considerable majority of Russia's politicians, parties, and movements signed it.

But the bungled intervention in Chechnya undermined the positive political trends of 1994. It is generally agreed that serious mistakes were made in the decision process to intervene and in the initial phases of the military operations. The intervention caused a break in relations between democrats—who opposed the intervention—and Yeltsin. Relations between the Center and the many diverse regions of Russia which appeared to stabilize in 1994 were undermined. Federal institutions, especially the armed forces, were weakened, and regionalism and the desire for autonomy and even independence from Moscow was reenergized.

Yeltsin's regime held to the course of reform in 1995 in substantial measure. But broad segments of political and popular opinion became more alienated from him and his policies. Because of this, the democrats' failure to unite, and Chernomyrdin's inability to forge an appealing centrist bloc, opposition communists and nationalists turned in a strong showing in the 1995 Duma elections, as they did in 1993. This election launched Russia into the most important, volatile, and uncertain phase of her politics since the collapse of the USSR: the run-up to presidential elections in June 1996. The constitution and circumstances make the Russian presidency supremely important, if not all powerful. Elections, Yeltsin's weak health, extra-constitutional moves by the "power structures," or some combination make a presidential transition over the new year highly likely. But the process and the results are very hard to predict.

This brief summary of political developments during the past several years says, in a word, that Russian politics have not settled down. Russia remains fundamentally unstable. The institutionalization of democratic life, the development of democratic political culture, party formation, founding the rule of law in laws, all are in their infancy. Russian political culture retains its special character: A propensity to personalization, authoritarianism, corruption, conspiracy, and confrontation and an aversion to coalition-building and compromise among the elites, and a propensity to apathy or anger among the population.

Few in Russia imagine that restoration of communist rule as it existed before the collapse is possible or desirable. But substantial elements of the political class and the population do believe that a much more authoritarian state and a much more state-controlled economy are desirable. At some point in the course of this revolution, these people and sentiments seem likely to get a chance to rule. Whether and when this could happen will depend primarily on developments on three fronts, which are, in order of importance, the economy, crime, and the military.

THE ECONOMY

The Russian economy is far from exiting the conditions of decline and turbulence that have been afflicting it now for several years. A difficult stage lies ahead as bankruptcies and/or payment arrearages close state enterprises and unemployment grows. What compensating factors, such as new jobs in the growing private sector and government welfare payments, will be available to ameliorate conditions, and how long before a real upturn are very uncertain at this point.

But, as the International Monetary Fund has recognized in agreeing to provide funds to finance the government deficit and help stabilize the ruble, there have been some positive developments in the last two years compared with the prior two years. Most important of these has been control of inflation. This achievement is largely thanks to government discipline in controlling emissions and subsidies. It has pleased international financial institutions, improved the atmosphere for investment, and eased the plight of consumers, especially those on fixed incomes. As a political statement, this achievement is the most tangible evidence so far that the management class still dominating Russian economic activity, of which Prime Minister Chernomyrdin is the senior representative, has become convinced that there is no turning back on the road to a market system. It remains a matter of which route to take and how to manage inevitable skids and bumps.

Official data and varied evidence from everyday life indicate that the slide in overall living conditions for the Russian population has slowed, may have halted, and might even be turning around. The government estimates that real incomes increased in 1994 by around 11 percent over 1993, but may have slipped again in 1995. The major factors here were the freeing of prices, the growing private sector in consumer goods and services, and imports. The consumer can get things simply unavailable before. The portion of the average income expended on food has dropped. The specter of widespread food shortages, even famine, seriously feared up to 1993 and 1994, has faded. These improvements coexist, however, with 30 to 40 percent of the population living at or below subsistence income levels and with growing income disparities that cause wide resentments.

It is not yet fully clear whether inflation control has really reversed the trend of declining investment in a country where the capital stock is woefully obsolete and the infrastructure deteriorating. Official data say investment is still declining. At the same time, the government recently asserted that foreign investment is on the rise, as are the numbers of joint ventures being formed and foreign business activity. Some reports claim that the flight of Russian native capital has slowed and may be starting to return to an environment in which quick profits are possible.

One overwhelming reality of the past several years has been the collapse of industrial production, down a massive 50 percent since 1989. Although many Russians blame this decline for their sorry economic plight and blame Yeltsin for it, it is worth noting that two-fifths of the drop occurred in the last years of Gorbachev because the old economic mechanisms were already falling apart. This, more than dogma about shock therapy, forced Yeltsin and Gaidar to suddenly free prices in 1992. Much, but not all, of this production drop has been economically and environmentally healthy as "value subtracting" enterprises—those producing unwanted goods, wasting resources in the process—ceased producing. Energy and raw material consumption has dropped accordingly, as has the generation of new pollution. Still, some of the drop has not been constructive, for it included, according to official data, energy sources vital to physical survival in some areas and to hard currency earnings, consumer goods and basic foodstuffs in a needy country, construction materials in a land desperate for additional housing (although the material for a visible private building boom has to be coming from *somewhere*), and investment in a country with antiquated capital stock and collapsing infrastructure. One of the puzzles of the Russian scene has been the apparent absence of serious analysis on how to distinguish between value adding and value subtracting industrial activity seemingly central to any government strategy of managed bankruptcy and restructuring. At the very least, the Chernomyrdin government promised a reformist industrial policy, but has yet to produce it It is unclear whether opposition elements, should they ever acquire real power, can produce one either.

The other, countervailing reality of the Russian economic scene is privatization and the rapid growth of private enterprise, especially in services, such as construction, trade, and financing. It is estimated that private activity now accounts for about 40 percent of Russian gross domestic product, is a major contributor to real personal income, and is a substantial buffer against unemployment. It is also, as is widely noted, impossible to measure accurately, because government statistical mechanisms are weak, and private enterprises have every incentive to underreport their activity.

In the past two years, more than 80,000 small businesses have appeared, and several hundred thousand private farms have come into existence. On the order of 5,000 state enterprises have offered shares to the public under the now completed voucher program. Many state firms are learning to operate in market conditions; negotiating their own supplier and customer relations, where these were previously decided at the center; trimming labor and other costs; etc.

Apart from macroeconomic conditions and crime, privatization is afflicted with two big problems. First, the government has not proceeded very far in creating the regime of transparent and consistent codes, laws, and regulations needed for a stable market economy to operate properly. Second, through various devices, the old management class—the nomenklatura, which ran things in the name of socialism—has seized the overwhelmingly dominant position in the new privatized enterprises to be run in the name of the market. This may be acceptable economically if the new/old management knows what it is doing. But it harbors an offense to social justice, as well as to the rule of law, that could ultimately harm marketization and democratization in Russia.

A more immediate problem is that this nomenklatura privatization often amounts to little more than plunder. Bosses of state enterprises privatize productive parts of their operations or set up private enterprises as satellites of state enterprises, using state supplied subsidies, materials, and technology on the input side while pocketing (and expatriating) profits on the output side. These are the very people who now dominate the Russian government. At the beginning of 1995, the question was whether they would seriously try to rescue the Russian economy, or, while buying off the population with increasingly worthless money, continue to plunder and cause real collapse and social upheaval.

The Russian economy, and with it Russian society, is at a crucial new stage that will be the next big test of Russian government and politics. The major facets of this situation are the payments or arrearages crisis among enterprises, inflation control, bankruptcy policies, and unemployment. To control inflation, the government withholds money. Enterprises in turn stop paying their workers and their suppliers. Enterprises are simply forced to shut down, often the efficient along with the wastrels. This problem has been nagging the Russian economy since the beginning of the reforms; it is now again severe. If the government funds the interenterprise debts, inflation will again shoot up (and many expect some such effect will have to be seen soon). It cannot do nothing. Some enterprises have to be sacrificed through a bankruptcy process, and some will be saved. In any case, enterprises will close, be liquidated, or be restructured. It will happen more or less spontaneously or through a kind of industrial policy applied through bankruptcy. The result will be increased unemployment. Real, as opposed to "registered," unemployment in Russia is now believed to be around 10 to 12 percent of the workforce.

Understanding this linkage, the Yeltsin-Chernomyrdin government is not going to simply bulldoze state enterprises over the cliff and let the unemployment chips fall as they may. It can be expected to try to manage emissions and subsidies, bankruptcy policies, and privatization to facilitate a soft landing. It is clearly now casting about for a strategy for doing this, and seems not yet to have found one. The essence of any such strategy, however, will be to control the rate of enterprise closings such that the growing private sector can absorb enough of the unemployed to avoid a social explosion. One aspect of the strategy will be privatization for cash. In the next phase of privatization, shares in privatizing enterprises will be sold for cash, rather than the state-issued vouchers of the first phase. This cash will be used, among other things, to underwrite new social safety nets through subsidies to local governments for that purpose. This makes the still chaotic center-region relationships newly important. The center depends on the regions for revenues, and the regions depend on the center for subsidies to smooth out the social, and political, consequences of regional differentials in wealth. This situation calls for patient and careful but determined management, not shock tactics. The question is whether the regime has the skill, discipline, and political control to maneuver through this minefield. Even if it does, it is clear much will depend on factors beyond its control. The growth of communist and nationalist opposition to reform in 1995, and the political pressures of a presidential campaign in 1996, will make this task even more difficult.

CRIME

Émigré sociologist Vladimir Shlyapentokh forecast some years ago that the criminalization of Russian society would be one of the consequences of the collapse of communism and would assume an influential role in Russia's development. Events seem to be proving him right. Eschewing colorful anecdotes and doleful statistics, this paper need only summarize the situation and draw some tentative conclusions.

The phenomenon of criminalization in Russian society has basically three dimensions.

Crime in the streets against ordinary citizens is clearly on the rise in Russia, mostly in the major cities. It remains well below the level of that experienced by residents of America's inner cities. But against a popular experience of and nostalgia for the order of the Soviet period, common crime, especially violent crime, fuels an appetite for order, discipline, and authority among the populace. They sense a society spinning out of control.

Organized crime, involving extortion and shockingly frequent violence, afflicts the business and political environment. This involves protection racketeering and frequent threats and acts of violence against businesses. As compared to organized crime in the West, the difference appears to be the lack of any "sectoral boundaries" beyond which the criminal element does not reach.

Corruption of officials and official institutions is endemic. It touches the police, customs and tax officials, privatization authorities, and the officials who control valuable state property and resources. Bribery and back-room transactions are commonplace. Well-connected industries—such as oil and gas, represented in the government by no less a figure than Prime Minister Chernomyrdin—have gained tremendous tax and other advantages, enabling their new owners to garner great riches, which are, often as not, kept in hard currency accounts in the West.

Russians and outside observers should understand that this is not just the product of postcommunist breakdown. Criminality in Russia has its roots in the Soviet experience of arbitrary authority, violence, and official corruption. The role of "friendships and connections" (*znakomstva i svyazi*, or ZIS, the initials of the Soviet Cadillac) was well established under Stalin. By the Brezhnev period, someone who did not steal was considered either deprived or deranged. It is not surprising that the end of the communist era was marked by any number of scandals and crimes, from the antics of Brezhnev's daughter to the expatriation of billions of dollars from the coffers of the CPSU by the KGB in 1990 and 1991. Morality and trust became commodities of importance only in the most familial or private (or criminal gang) relationships. With the collapse of communist power, this thriving, but somewhat hidden, fungus simply exploded into the open.

There is, no doubt, some macroeconomic tax levied on Russian development by crime. Everyone in business must simply pay up the required 10 or 20 percent of profits to stay in business. That there are some natural limits to this is suggested by increased calls from established criminal elements that newcomers and interlopers should be brought to heel—an old story. There is clearly a political tax in the sense that criminality and official corruption undermine the authority of government leaders and institutions.

Organized crime is now credibly established to be a factor in the linkages from Russian laboratories and military bases to foreign arms importers that make Russia a dangerous source of proliferation of weapons, materials, and technology for mass destruction. If there is a military threat emanating from Russia following the Cold War, this one is near or at the top of the list.

The gravest penalties entailed on Russia by crime and corruption, however, are likely to be social and psychological over the long term. After 70 years of Soviet corruption and 10 or 20 years of postcommunist corruption, how will emerging generations develop a sense of law and order, contract, and mutual obligation? Russia may evolve a more or less democratic and market-based society. But given its size and potential power, it matters to all whether that society looks, in terms of the power possessed by the law, more like Sweden or more like Sicily.

In any case, among the major items on the agenda of the Russian government is combating the bacillus of crime with which it is itself infected. Showing some results is probably as important as progress on economic stabilization and recovery and is closely related.

THE RUSSIAN MILITARY

The military presents Russia with two problems. The first is to decide on a defense policy: What force posture, weapon programs, doctrines, organizational reforms, manning policies, etc., are needed to meet Russia's security requirements commensurate with its new post-Soviet domestic and international situation? Of course, no country in a state of revolutionary change can decide and implement such matters at a stroke. This gives all current plans and pronouncements a very transient quality, although Russian military policy discussions do provide insight into the kind of power Russia will try to be in the years ahead.

Getting military policy sorted out is of great importance to Russia's defense leadership, because that is their business, and there are large economic and social implications from what they try to implement. Nevertheless, at present the implications for Russia's basic security are not grave. This may sound strange given the turbulent and, in places, violent surroundings of the so-called near abroad, the former republics of the USSR. But the present period of upheaval is unlike many in Russian history. Often in the past, when Russia underwent revolutionary change or internal fragmentation, it was vulnerable and subjected to outright invasion by neighboring powers of comparable military strength, such as Poland, Germany, and Japan. The very survival of the Russian state could be threatened from abroad. Today this is not the case. All of Russia's past or potential great-power adversaries lack the interest or the capability to interfere in its revolution in ways that would threaten its survival, or even much influence the outcome. This is largely for political reasons; in other words, even if Russia were militarily prostrate, neither America, nor Germany, nor Japan, nor even China, not to mention such other troublemakers of the past as Sweden, Poland, Lithuania, or Mongolia, has the remotest intention of attacking it. This could conceivably change in the case of China, but seems unlikely to do so for the foreseeable future. For Russia has inherited the bulk of the USSR's strategic nuclear arsenal. For once, Russia can have its revolution in peace, save for the threat of internal violence, which is its business, and violence on its periphery, which it has ample power to prevent from endangering its survival.

Some sophisticated Russian analysts, such as Sergei Karaganov, appreciate this explicitly. Traditional paranoia and xenophobia persist in the population at large, in the military, and among extreme nationalists. But, over time, the absence of clear and present great-power threats to Russia, whatever tensions and rivalries may arise, should tend to mellow those attitudes, demilitarize Russia's sense of its statehood, and contribute to healthy political outcomes.

Russian debates about military policy and doctrine reflect many different views, but there seems to be a core consensus implicitly accepting the threat picture sketched above. Russia should remain (or return to being) a superpower, whatever that is in the post–Cold War world. It should maintain a robust nuclear deterrent—and as conventional capabilities deteriorate, Russian reliance on its nuclear weapons has increased. Russia has adopted the posture of first use of nuclear weapons. For the future, Russia would like its general-purpose forces to be characterized by high technological quality, mobility, and flexibility, but to be much smaller than those of the USSR. For the near term, Russia appears to be seeking to dominate the area of the former Soviet Union—the so-called near abroad. For the longer term, the Russian military would like to have forces capable of power projection over longer distances.

In the meantime, Russia faces a second problem in the military: organizational decay, alienation, and demoralization. These problems increased with the invasion of Chechnya. The military performed poorly and suffered many casualties. Because of these problems, the military could alter the country's political and economic development through coups or mutinies; however, this is much less likely than corporate activism in politics or a failure to defend legitimate civilian authority. Although the military were active in the latest Duma campaign, neither of these threats clearly emerged. Yet they remain a danger for Russia.

In 1988, the armed forces of the USSR numbered 4.5 million people. The breakup of the USSR, demobilization, desertions, and conscription shortfalls leave Russia's armed forces at about 1.5 million today. Russia has now almost fully withdrawn military forces from the former Warsaw Pact countries and the Baltic states. With the exception of nuclear forces in Ukraine, Belarus, and Kazakhstan (being withdrawn under the January 1994 agreement); the Black Sea Fleet (whose division with the Ukraine is still being negotiated); the 14th Army in transdniestrian Moldova (an exceptional situation itself); and relatively small Russian "peacekeeping" contingents in Tadzhikistan and the Transcaucasus, the remainders of Soviet military forces now located in the former republics are nominally under republic command. Russia's ground and air force units number about 50 percent of those of the USSR. Its navy has shrunk overall about 40 percent. Significantly, its SSBN force remains about the same size; retaining it is a high priority for Russia because it provides strategic deterrence and superpower status. It is also important for the Russian navy because it assures institutional survival in a land-force dominated military.

Russian defense procurement has shrunk to about 20 percent of Soviet levels in 1989. Weapon modernization has slowed to a crawl, but not stopped entirely. Most Russian defense spending is now devoted to paying, feeding, and housing military personnel and that at vividly inadequate levels. The USSR's vast military industrial and R&D establishment has fragmented and is now rapidly shrinking in Russia, as well as the republics, because of low procurement, payment arrearages, and some conversion to civilian production. Domestic military needs and foreign sales will not sustain anything like its former size.

The demoralization and alienation of the Russian military has many causes and manifestations. It has lost much of the exalted status enjoyed since 1945, although it is more popular than most other institutions in the society. Even before the collapse of the USSR, the military's image and self-esteem suffered from Afghanistan, dismal everyday living conditions for conscripts, and general awareness that military spending had bankrupted the economy. Features of today's

Russian armed forces that make it a severe social and a potential political problem are

- low morale and discipline

- corruption, e.g., theft, bribery, misuse of military property

- resentment of the civilian leadership (in significant measure because of the military's having been drawn into the events of October 1993 and Chechnya)

- growing divisions among ranks and generations in a now very top-heavy personnel structure (disgruntlement among the field grades is particularly severe because they sense they have few career prospects and do not enjoy the privileges of general officers)

- geographic divisions as units become more dependent on localities for their maintenance and sense of role (General Lebed's 14th Army in Moldova was a case in which this phenomenon has reached the level of local warlordism)

- lack of respect personally for the president and the minister of defense—especially after Chechnya

- a political outlook that is increasingly antidemocratic, nationalist, and anti-West.

What kind of political behavior is to be expected from the Russian military? Despite the troubling developments sketched above, the most likely behavior is continued disgruntled suffering along with the rest of Russian society. In day-to-day politics and during the run-up to parliamentary and presidential elections during the coming years, the military is likely to support nationalist and authoritarian figures and programs. It may well even produce a major authoritarian candidate, such as Lebed. A coup or some other corporate intervention in politics is possible, but not very likely unless there is a breakdown of the government. Divisions within the military are themselves obstacles to corporate political intervention on the initiative of military leaders. Should a showdown like that of October 1993 recur, the military is more likely to stand aside or, if it acts, to support antireform elements. Under such conditions, the likelihood of the military breaking apart is arguably high. In sum, the political role of the Russian military in the years ahead probably depends more than anything else on the political responsibility and sobriety of the politicians.

SECURITY ISSUES

After an initial phase of what its critics called a "romantic" pro-Western orientation, Russia is seeking to forge a more assertive foreign policy that is distinctively Russian and based on a more nationalistic interpretation of Russian interests. This has occasioned frictions and given rise to nationalistic official rhetoric from President Yeltsin and Foreign Minister Kozyrev, no doubt in part to appease domestic nationalism and to defuse the opposition. But in practice, and largely for economic reasons, Russia is trying to improve relations with all significant countries and regions. This itself may cause frictions, for example, with the United States, as Russia struggles to increase arms and reactor sales in Southwest Asia.

Nevertheless, four major subjects are sure to trouble Russia's relations with the United States and the West over the coming years: Russia's actions in the near abroad, its role in official and criminal arms trafficking and reactor sales, its posture on European security and NATO expansion, and its posture on strategic offensive and defensive systems.

Instability, and in some cases, ungovernability within and among the former Soviet republics constitute Russia's most pressing security problem. A policy of total detachment is not a practical option, for either foreign or domestic reasons. At the same time, most Russians recognize that trying simply to reassemble the USSR or the old Russian empire is also not viable. What is emerging is a policy of differentiated and selective intervention, with qualified economic and military reintegration (via the CIS). Russia's basic interests are

- to contain violence and instability that might escalate and spill into Russia

- to prevent intrusion into its periphery of significant foreign influence

- to protect Russian populations in the new states from abuse or expulsion

- to organize commonwealth-type relations where possible in ways that enhance Russia's influence and dominate the region geopolitically without excessive economic cost.

Russia's perceptions, interests, and policies recognizably, although not officially, divide the near abroad into three regions that are treated differently but in line with the above principles.

The first category is the Baltic states. Border problems and the Russian minorities will remain significant sources of friction. But here Russia has gone the furthest in not only recognizing officially (as in all other cases) but accepting psychologically and politically the fact of their independence. Diplomacy and perhaps economic levers will be the means for protecting Russia's interests here. Russia recognizes that interventionism and strong-arm tactics are most likely to offend and arouse the West in the Baltics, at costs Russia cannot afford and need not pay to protect its interests there. As of late 1995, there have been ominous exceptions to this good sense expressed by the communist/nationalist opposition, ostensibly in response to plans for NATO expansion. Whether such "neo-imperialist" views influence official policy must be a cause for concern.

The second category spans the Transcaucasus and Central Asia (Uzbekistan, Turkmenistan, Kirgizstan, and Tadzhikistan). These regions are regarded as buffers needing active defense. Russia wants to be present militarily in areas near the borders of the old Soviet Union, and is ready to use military force and to intervene in local politics to maintain Russian influence and to exclude significant foreign influence. Here is where Russian behavior is most likely to appear imperialistic to the West or seemingly inimical to democratic nation-building. In the case of oil-rich Azerbaijan, Uzbekistan, and Turkmenistan, it will seek to derive economic benefits by exploiting their need for transit through Russia. Russia does not oppose democratic state-building in the south but is strongly inclined to see what might look like democratic politics as an invitation to "hostile," e.g., Islamic, intrusion. The odds favor Russia's being able to defend its essential interests in these regions on terms the West can accept if not always applaud. In the longer run, Russia and China might find themselves competing and/or cooperating over the region, depending on what happens in China and in China's Central Asian regions.

The third category embraces Belarus, Moldova, Ukraine, and Kazakhstan. These countries contain large Russian populations or are almost wholly populated by Slavic peoples culturally and historically close to Russia. Official Russian policy recognizes their independence. But Russian behavior in the future is very likely to be guided by the belief that all or large parts of these countries ought to gravitate back into Russia. Russian policy will prefer this reassimilation to be gradual, consensual, and peaceful, to avoid trouble for Russia locally or from the West. Geography and ethnography, however, auger against a tranquil scenario. An all too likely alternative scenario, to generalize about complex circumstances, is that Russian or pro-Russian elements will agitate for secession and reassimilation, causing a split of the country and intrastate conflict into which Moscow must intervene. Ukraine and Kazakhstan are the key cases. Depending on the circumstances, Russia may appear as an imperialist or as a defender of democratic principles and human rights. But any such scenario is bound to cause tensions with the West.

The issues of arms trafficking and the sale of reactors are unavoidably on the agenda of Russia's relations with the United States. Russia seeks through official foreign military sales and the sale of nuclear power reactors to such countries as Iran to earn hard currency, to keep vital industrial and research capabilities intact, and to show the flag for influence and prestige. The fact that the plausible scale of such trade will be limited by customers' inability to pay, noncompetitiveness of many Russian offerings, and deficiencies in Russia's ability to support advanced systems will not deter an energetic sales effort, although it may mean that the final results will be meager. As the reactor deal with Iran indicates, Russia is becoming less sensitive to American and other expressions of concern about its foreign nuclear reactor and arms and weapons technology sales. Russia wants the sanctions on Iraq lifted in order to sell weapons to Baghdad.

In addition, the United States and the world are increasingly likely to face another problem from Russia: an unofficial (but officially tolerated) and illegal (and uncontrolled) traffic in arms, technology, and talent. It is hard to imagine any Russian government that would not want to prevent the uncontrolled expatriation of capabilities related directly to weapons of mass destruction, i.e., nuclear, chemical, and biological weapons. At the same time, it is hard to imagine any

Russian regime in the next decade being so fully in control of the relevant resources and actors that it can assuredly prevent such transfers. Economic incentives, permissive disorder, and enabling government corruption are simply not going to be eradicated soon. An examination of the details of this market and of recent illustrative episodes is beyond this chapter. Suffice it to say that the problem will be around for some time. Dangerous proliferation out of Russia will continue, probably getting worse before it gets better. It will be the occasion for constructive cooperation to stem it, as well as frictions arising from Russia's seeking to avoid politically costly blame or politically unwelcome foreign intrusion into its sensitive affairs. The main political problem will be Western perceptions that Russian official entities are, at least partly, culpable and that the Russian government is unable or unwilling to do anything serious about it.

In coming years, something different from what obtains today will have to be arranged for the security of the countries between the Baltic, the Balkans, and the Black Sea. The foremost option presented so far is the expansion of NATO eastward. Despite joining the Partnership for Peace, Russia has made clear that any expansion of NATO that excluded Russia would be intolerable (it would be seen as a dire threat by nationalists and as a political danger by reformers because Russia would be isolated), that any expansion of NATO would be undesirable, and that the beefing up of the Organization of Security and Cooperation in Europe as a collective security mechanism, and the transformation of NATO within this mechanism, would be preferable. Russia's political development, democratic or otherwise, should influence its position on this issue. In theory, a stable democratic Russia should have no problem with East European neighbors joining an alliance that gives them a sense of security but does not threaten Russia. But, for now, Russia bitterly opposes NATO expansion—although it does not appear to be able to prevent it. Should expansion occur, the current trends in Russian policy point in the direction of increased Russian pressure on the Baltic states and Ukraine, and of the expansion of ties with Iran and others.

Because of its increased emphasis on nuclear weapons, and in keeping with its generally more assertive policy, Russia appears to be unwilling to ratify the START II agreement in its current form. Russia is also reported in press accounts to be protecting its biological weapon program and continues to oppose an American ballistic missile defense capability.

SOME FORECASTS AND IMPLICATIONS

If this assessment is valid, one must conclude that Russia's political future is highly uncertain, sure to be a generation or more in working out, probably marked by unpleasant detours toward authoritarianism and away from economic reform, and possibly seeing fragmentation and widespread civil conflict. At the end of this process—which could take a very long time—some form of integrated and even powerful Russia is likely to reemerge. Whether we see history's familiar result, a Russia that is autocratic, xenophobic, and backward, but powerful enough to be threatening, or a stable, more or less democratic Russia, powerful economically and strong enough for its own needs militarily, but not threatening—that is the historic question. We shall simply have to live a long time without a final answer.

Should a swerve to authoritarianism take place in the next few years, it is likely to take the form of a nationalist, quasi-communist reversion and prove transient because of its inability to govern and revive the country. It could even precipitate fragmentation and civil war. In the longer term, authoritarian alternatives could emerge that, because of more enlightened policies, particularly on the economy, would have a better chance of surviving and of reviving Russia. They might also have a better chance of ultimately evolving into more democratic forms. The point of these speculations is that the timing of political zigzags is important for their consequences. The further Russia develops toward an economy and a society independent of the state, the more likely it is that political currents will stabilize into democratic channels, although some authoritarian alternative should probably not be totally excluded from the politics of a socially advanced and economically capitalistic Russia.

All in all, however, Russia's current difficulties imply that current efforts directed toward nuclear "risk reduction" remain vital and may even take on added significance if the country becomes more chaotic. It is even possible, although very unlikely, that we could become involved in a peacekeeping operation on the territory of the former Soviet Union. In the long term, the possibility of a revived and antagonistic Russia implies that we must retain the ability to reconstitute our forces if such a threat should emerge. The odds favor a democratic outcome in the very long term more than ever in the past, and this is not because of developments in current Russian politics. Such a judgment rests, rather, on more fundamental realities: an educated population, a society that cannot be closed off from the world the way Russia and the Soviet Union were in the past, and a national desire for a powerful and prosperous nation that cannot exist unless it masters modern technology, invites foreign investment, and interacts vigorously with the world.

BIBLIOGRAPHY

Yergin, Daniel, *Russia 2010 and What it Means for the West*, New York: Random House, 1995.

Chapter Four

THE BALKANS

F. Stephen Larrabee

The Balkans have traditionally been a source of tension and instability in European politics. Deep-seated ethnic feuds and territorial disputes have earned the region a reputation as the "powder keg" of Europe. Many of these disputes were suppressed or lay dormant during the Cold War. However, the collapse of communism has revived many of these conflicts and has given them new intensity (Larrabee, 1994). This has hindered the emergence of a stable security order in the region and has thrust the Balkans back to center stage in international politics.

Indeed, the Balkans are becoming Europe's new front line. It is here, rather than in East Central Europe, that the new conflicts on Europe's periphery are most acute, and it is here that Western interests have become most directly engaged. The Balkans have also become the testing ground for possibilities—and the limits—of the West's effort to develop a new cooperative security relationship with Russia. Hence, the Balkans are likely to remain an enduring Western concern long after the ink has dried on the peace accord signed in Dayton.

DAYTON AND BEYOND

Much will depend on what happens in Bosnia and how well the Dayton agreement holds up. It could still fall apart. The main danger is that the Bosnian Serbs will simply lie low and wait until U.S. troops withdraw in a year or so, then restart the fighting. There is also danger of a "second Somalia"—that U.S. troops sustain a few well-dramatized casualties, which leads to growing domestic pressure to bring the U.S. troops home early. The impact of such a move on regional stability—and Alliance relations—could be quite severe.

But even if the Dayton agreement holds, the Balkans are likely to remain characterized by continued instability and unrest. With the exception of Greece and, to a lesser extent, Turkey, all the countries in the region lack strong democratic traditions and institutions. Many are plagued by unresolved minority problems. Hence, their transition to stable democracies is likely to be difficult. Moreover, the Dayton agreement has left two of the most important problems in the region—Macedonia and Kosovo—virtually untouched. Without a satisfactory settlement of these issues, there is not likely to be lasting peace and stability in the Balkans.

Macedonia (FYROM)

The situation in Macedonia is potentially quite unstable. In the last two years, tensions between the Slav population and the Albanian minority, which comprises 23 percent of the population, have escalated.[1] These tensions came to head in February 1995, when the Albanian community sought to establish an Albanian-language university in Tetovo, the main Albanian town in Western Macedonia. Clashes broke out with the Macedonian police that left one Albanian dead and some 60 others wounded.

Moreover, the United Nations embargo against Serbia, together with the embargo imposed by Greece against Macedonia in February 1994, has resulted in a serious deterioration of the Macedonian economy. Industrial production dropped significantly in 1995 over 1994. Inflation is over 50 percent. As a result of the two embargoes, moreover, corruption and smuggling have reached epidemic proportions.

A stabilization of the situation in Macedonia is a prerequisite for a stabilization of the Balkans as a whole. If Macedonia explodes, it could ignite the entire southern Balkans, bringing in Albania, Serbia,

---

[1]For background, see Glenny (1995). Also see Perry (1995).

and Bulgaria—all of which have historical claims on Macedonia—as well as Greece and Turkey.

The United States, which has some 500 peacekeeping troops in Macedonia, would be affected by any instability there. The Clinton administration (or its successor) could be faced with strong congressional pressure to withdraw these troops to avoid another "Somalia"—a move that would have broader implications for stability in the Balkans, as well as for U.S.-European relations generally. Hence, the United States and its European allies have a strong interest in defusing ethnic tensions in Macedonia and preventing an internal explosion there.

Future developments in Macedonia will be affected by several factors. The first is the Macedonian government's handling of the concerns of the Albanian minority. These center primarily around demands for greater university instruction in the Albanian language. The Macedonian authorities fear that the establishment of an Albanian university at Tetovo is the first step toward setting up parallel institutions, such as those in Kosovo (see below), and eventual separatism. In the aftermath of the clashes at Tetovo, the government took steps to increase the opportunities for Albanian instruction at the University of Skopje. This has helped to defuse tensions somewhat, but the issue still remains potentially explosive.

The second important factor will be Albania's attitude and policy. To date, Albanian President Sali Berisha has played a moderating role in the conflict between the Albanian community and the Macedonian government. This has helped to keep tensions from further escalating. However, if this policy were to change, unrest in Macedonia could increase. Thus, it will be important for Western governments, especially the United States, to continue to press Berisha and other Albanian leaders not to stoke the fires of Albanian nationalism, either in Kosovo or Macedonia.

The third important factor will be Greek policy. Much will depend on how Greek-Macedonian relations develop in the next few years and whether Greece and Macedonia can succeed in settling their differences, particularly over the name of the Macedonian state. The interim agreement signed on September 13, 1995 is an important step in this direction. It calls for mutual recognition, opening trade routes, and establishing liaison offices in Athens and Skopje. As part of the agreement, Greece agreed to lift the embargo imposed in February 1994. In return, Macedonia agreed to drop the star of Vergina, regarded by Greeks as an important symbol of Greek culture, from its flag and to alter a number of provisions in its constitution that Greeks claimed implied that Macedonia harbored territorial ambitions against Greece.

The agreement has helped to defuse tensions between the two countries and lays the basis for a more far-reaching reconciliation over the long run, including a resolution of differences over the name of the Macedonian state (the main unresolved issue). This, in turn, could result in a mutually beneficial expansion of trade and other relations. Greece and Macedonia are, in fact, natural partners. Greece is the only one of Macedonia's neighbors that does not have territorial claims against Macedonia. It is also a natural economic partner. Closer economic ties with Greece would enable Macedonia to reduce its economic dependence on Serbia and rebuild its shattered economy. It would also give Macedonia a much-needed and much-desired outlet to the sea (Salonika) for its agricultural and manufactured products.

Closer economic ties with Macedonia would also have significant benefits for Greece. They would give Greece important access to the Macedonian market and create a network of economic ties that over time could make political differences easier to resolve. Indeed, if Greece and Macedonia could resolve their political differences, Salonika could become the economic and financial capital of the Balkans, making Greece the dominant economic power in the region.

Kosovo

The other major problem that still needs to be resolved is Kosovo. Kosovo remains the Achilles heel of the rump Yugoslav federation. The province, whose population is 90-percent Albanian, has been a persistent source of instability and nationalist unrest since the late 1960s. Riots erupted in 1968 when Albanian students demanded the establishment of an Albanian-language university and the granting of full republic status to Kosovo. Student protest erupted again in

1981, one year after Tito's death. At that time, demands were also raised that Kosovo be made a republic.

The nationalist upsurge in Kosovo stimulated a backlash among the Serbs, which Serbian President Slobodan Milosevic consciously exploited in his rise to power. In 1989, he stripped Kosovo of its autonomous status and sharply curtailed the rights of the Albanian population. Since then, Kosovo has been under virtual Serbian occupation. Thousands of Albanians have lost their jobs, and several hundred have been arrested.

At the moment, the situation in Kosovo is quiet. The Albanian population has adopted a policy of passive resistance and has established a parallel state structure, with its own constitution, hospitals, schools, and administrative structure. At the same time, the Kosovars have sought to "internationalize" the problem and call attention to their lack of political rights. However, Kosovo remains a potential flashpoint. Unless a settlement is found that restores political and civil rights to the Albanian-speaking population, there is a danger that some incident, either accidental or provoked, could lead to the outbreak of large-scale unrest in Kosovo.

Any unrest among the Albanian-speaking population in Kosovo would heighten tensions between Serbia and Albania and could spill over into Macedonia, which, as noted, has a large Albanian population. Thus, internal stabilities in Kosovo and Macedonia are closely linked. Moreover, both President Bush and President Clinton have officially warned the Serb government that the United States would not tolerate a Serb crackdown in Kosovo and that it would take action to prevent it. Thus, U.S. credibility and interests are also involved.

The Dayton Accord has left a sense of bitterness and resentment in Kosovo. Many Kosovars had hoped that a Bosnian settlement would also include provisions to restore their political rights. The fact that it has not has left many Kosovars feeling deeply disappointed. Moreover, the degree of autonomy granted the Serbs in Bosnia sets an important precedent. Many Kosovars may now press for a similar arrangement and see it as a possible stepping stone to eventual unification with Albania.

At the same time, the prospects of reaching a settlement of the Kosovo issue may be better now than before the conclusion of the Dayton accord. With Bosnia resolved—assuming the accord holds— the Kosovo issue is likely to move to the top of the Balkan policy agenda. Milosevic knows that he has little chance of getting Western assistance to rebuild his ravaged economy unless he addresses the Kosovo issue. Western businesses, moreover, want stability. They are unlikely to invest heavily in Serbia if the situation in Kosovo remains unstable and threatens to erupt.

Over the long run, moreover, demographic pressures are likely to increasingly push the Kosovo-Albanian issue to the forefront. According to some estimates, if current demographic trends continue, some 18 million Albanians will live in the Balkans by the second or third decade of the 21st century (Glenny, 1995, p. 27). This means that the Serbs and Macedonians will need to find ways of accommodating Albanian political aspirations or face the prospect of large-scale political turmoil and the possible break-up of their respective states. Indeed, the problem of how to accommodate Albanian nationalism and national aspirations is likely to become an increasingly important issue in the Balkans in the years to come.

Both these factors give Milosevic an incentive to resolve the Kosovo issue sooner rather than later. The longer Milosevic waits, the more numerous and stronger the Albanians will become. Moreover, it may be easier for Milosevic to strike a deal now—while he is strong and faces relatively little internal opposition—rather than later, when pressures for internal democratization within Serbia may be greater and his own bargaining power has been eroded.

Indeed, the real danger of an explosion in Kosovo is not in the next six months or year, but in several years when pressures for democratization *within Serbia* are likely to be stronger than they presently are. The pressures for greater liberalization within Serbia could encourage the Kosovars to escalate their demands and press harder for full secession rather than greater autonomy within a reorganized federal Yugoslavia. Most revolutions or outbreaks of large-scale social unrest occur not at times of great repression but during periods of liberalization when the reins of central power have been weakened and expectations are rising—as the recent histories of the Soviet Union under Gorbachev and of Yugoslavia itself underscore.

Hence, the time to strike a deal in Kosovo is now, rather than later. The longer Milosevic waits, the harder it will be.

Croatia

Croatia has emerged as an important winner from the Bosnian conflict. It has not only recaptured much of Serb-occupied territory in Krajina, expelling most of the Serb population in the process, but also regained Eastern Slavonia as well. It may also obtain much of Herzegovina, thus laying the basis for the creation of a Greater Croatia. This has long been a cherished goal of many Croats, including President Tudjman himself. Indeed, one of the main problems in the future may be containing Croatian nationalism and territorial aspirations.

Tudjman wants to tie Croatia more closely to Europe and the West. He cannot do this, however, without Western support. This gives the United States and its European allies, especially Germany, a certain degree of leverage, which they may be able to use to encourage Croatia's domestic liberalization—including respect for minority rights—as well as its adherence to European political, economic, and social norms.

Serbia

Regardless of the outcome of the Bosnian conflict, Serbia will remain an important actor in the Balkans. The key question is: What type of actor? Will it be an aggressive, nationalist Serbia, which poses a threat to its neighbors, or a benign, democratic Serbia, ready to play a constructive role in building security and stability in the Balkans? At this point, the answer is not clear. Much will depend on Milosevic's own goals.

A policy of ostracism and exclusion, however, is not likely to work over the long run. It will simply reinforce Serbia's historical penchant—clearly evident during the Bosnian conflict—to see itself as a "victim" and to play into the hands of Serbian nationalists and extremists. It is also likely to make Serbia ally itself with and seek political support from like-minded pariah states that share its nationalist ambitions. This, in turn, is likely to make Serbia a continuing source of instability in the Balkans and a potential threat to its neighbors. Thus, the United States and its allies need to find a way to encourage Serbia's long-term democratic evolution and eventual reintegration into a broader European and regional framework.

BULGARIA, ROMANIA, AND ALBANIA

Security in the Balkans will also be directly affected by how well the other non–former Yugoslav postcommunist countries—Bulgaria, Romania, and Albania—manage their transitions. The pace of reform in these countries has been much slower than in East Central Europe. In large part this is because democratic traditions and civil society were weaker in the Balkans than in East Central Europe. Communist rule was also more repressive in the three Balkan countries than in East Central Europe. Thus, there were few organized independent groups that could fill the political vacuum created when communism collapsed. As a result, in the first round of democratic elections in 1990, postcommunist or neocommunist parties succeeded in returning to power in all three countries.[2]

The political situation, however, has evolved considerably in all three countries since then. In Bulgaria, the democratic opposition, centered around the United Democratic Front (UDF), a loose umbrella movement composed of various opposition groups, succeeded in winning the second round of elections in October 1991. However, the UDF soon split into various feuding factions and squandered the momentum and support it had built up, paving the way for the victory of the Bulgarian Socialist Party (BSP—the former communist party) in the December 1994 elections.

The strong showing of the Bulgarian Socialists in the December 1994 elections, however, reflected less a desire to see a return to communism than widespread public disenchantment with the impact of reform and the constant infighting within the UDF, which had seriously impeded the ability of the UDF to carry out a coherent program of economic reform. The BSP reaped the rewards of the UDF's mistakes and incompetence. The return to power of the socialists has

---

[2]Here, *postcommunist* refers to former communist parties that have regained power.

not resulted in a significant shift away from reform. However, reform, especially privatization, had not really proceeded very far. Today, only about 20 percent of the Bulgarian economy is in private hands, in comparison to over 60 percent in Poland and Hungary and about 80 percent in the Czech Republic. Inflation—about 65 percent—also is considerably higher than in East Central Europe.

On the foreign-policy front, the Socialist government has generally continued to pursue a policy of Western integration. However, on NATO, differences have emerged between President Zhelev—a strong advocate of NATO membership—and the government, which has adopted a more equivocal position and has tended to show much greater sensitivity to Russian concerns. In addition, the BSP government has pursued ties to Turkey with less ardor than its UDF predecessor.

Romania's transition has also been much slower than those in East Central Europe. President Ion Iliescu, a reform communist who fell out of favor with Ceausescu during the latter years of the late dictator's rule, has more in common with Gorbachev than with former East European dissidents, such as Vaclav Havel, Lech Walesa, or President Zhelev in Bulgaria. This has led many Western observers to dismiss Romania as a neocommunist backwater.

In the first few years after the collapse of communism, this view was perhaps justified. But Romania has evolved significantly since then. After a slow start, economic reform has begun to pick up speed, sparked by a revived large-scale privatization campaign. Today, the private sector accounts for 40 percent of gross domestic product— considerably more than in Bulgaria, but significantly below the level in East Central Europe. Inflation, while high by East Central European standards, has fallen significantly this past year.

The transition, however, has been painful. Living standards—the lowest in Eastern Europe during the communist period (with the exception of Albania)—have dropped below those of the Ceausescu era. Real wages are about one-third lower than in 1990. Romania also witnessed an upsurge in labor unrest in 1995 that suggests that the population's patience may be wearing thin.

In the foreign-policy area as well, Romania's position has evolved significantly. In the initial period after Ceausescu's ouster, Romania flirted briefly with neutralism, but it has embarked firmly on a pro-Western course since 1992. In 1992, together with Bulgaria, it signed an Association Agreement with the European Union (EU), and it is strongly committed to NATO membership at the earliest possible date. However, Romania's differences with Hungary over the treatment of the Hungarian minority represent an important obstacle to Romania's entry into either organization.

Here too, however, there has been a shift in Romanian policy lately. In September 1995, Iliescu launched an initiative aimed at achieving a "historic reconciliation" with Hungary along the lines of the French-German rapprochement (Rüb, 1995). Iliescu's initiative has helped to ease tensions with Hungary and has significantly improved the prospects for the conclusion of a long-delayed bilateral state treaty, which is designed to provide a framework for the normalization of relations between the two countries.

Iliescu has recently also sought to curb the influence of the extreme nationalists within the ruling coalition. In October 1995, the extreme nationalist and anti-Semitic Greater Romania Party was expelled from the ruling coalition.[3] The leader of the Greater Romania Party, Cornel Vadim Tudor, a former Ceausescu loyalist, had long been a thorn in Iliescu's side due to his extremist positions and personal attacks on Iliescu. The ouster of the Greater Romania Party from the coalition frees Iliescu to pursue a more moderate course domestically, as well as to continue rapprochement with Hungary.

Both moves appear to be part of a broader strategy to improve Romania's image in the West and to enhance its prospects to obtain membership in the EU and NATO. They reflect the growing recognition within the Romanian leadership that Romania has little chance of joining either organization unless it improves its political image and its record on minority rights. To be sure, Romania still has a long way to go before it is ready for membership in either organization. Nevertheless, the desire for membership in NATO and the EU has provided an important incentive for the recent efforts by Iliescu to speed up domestic reform and to regulate Romania's outstanding differences with Hungary.

---

[3]See "Bukarest bemüht sich um Image-Korreckur" (1995).

Albania's transition has been much slower and more problematic than the transitions in Romania and Bulgaria. But Albania also started from a lower base. It had the lowest standard of living of any country in communist Eastern Europe. It was also the most isolated and cut off from outside developments, including those within the communist world. Thus, overcoming the communist legacy and building a democratic system and market economy have proven to be even more daunting tasks than elsewhere in the Balkans and East Central Europe.

After winning a resounding victory in the March 1992 elections, the Democratic Party, headed by President Sali Berisha, has been plagued by infighting, inhibiting the implementation of a coherent reform program. Constitutional reform remains stalled. Berisha has also shown increasing authoritarian tendencies lately. If these continue, they could jeopardize Western support and economic assistance.

On the foreign policy side, the picture has been more encouraging. Relations with Greece have been marred by differences over the treatment of the Greek minority. However, after a sharp deterioration in 1994, relations improved in 1995 thanks in particular to American behind-the-scenes diplomatic intervention. Relations with Turkey have also been strengthened, especially in the military sphere.

However, the key issue for Albania—and for future stability in the Balkans—is the Kosovo issue. The Albanian government has a strong interest in the fate and treatment of the Albanian population in Kosovo whose political and cultural rights have been suppressed by the Serbs. To his credit, Berisha has—so far—not sought to stoke the fires of Albanian nationalism in Kosovo or among the Albanian community in Macedonia. However, as long as the issue of the rights of the Albanian-speaking population remains unresolved, there is always a danger that Berisha or some other Albanian leader may seek to play the nationalist card in an effort to increase his popularity or legitimacy, as both Milosevic and Tudjman did in Serbia and Croatia. Hence, a restoration of the political rights of the Albanian population in Kosovo and an effort to ensure respect for minority rights in Macedonia are important prerequisites for long-term stability and security in the Balkans.

GREECE: THE WINDS OF CHANGE

Stability and security in the Balkans will also be significantly affected by the policy that Greece pursues in the region. Greece has the potential to play an important stabilizing role in the Balkans. It is the most stable state in the Balkans politically; it has the strongest economy in the region; and it is a member of the EU, West European Union, and NATO. Thus, in comparison to the rest of the impoverished states of the Balkans, Greece looks like an island of stability and prosperity.

Greece, however, has played a good hand badly—at least until recently. During the 1970s and early 1980s, Greece had been one of the foremost champions of Balkan détente and had made an improvement of relations with its Balkan neighbors a cornerstone of its Balkan policy. The disintegration of the former Yugoslavia, however, heightened Greece's sense of insecurity and vulnerability. In effect, Greece saw itself as besieged on all sides, with no potential allies except for Serbia. This siege mentality was reflected in particular in the Greek obsession with Macedonia (FYROM) and prevented Greece from exploiting many diplomatic opportunities to contribute to greater stability in the Balkans. Instead of playing a stabilizing role in the region—a role to which it aspired and was potentially well-suited—it pursued self-defeating policies that put it at odds not only with most of its Balkan neighbors, but also with its partners within the EU. The end result was Greece's isolation in the Balkans and in Europe.

However, since early 1995, there has been a visible shift in Greek policy in a more moderate and pragmatic direction. The nationalist tide that characterized Greek politics from 1991 through 1993 has ebbed. A growing number of politicians on both sides of the Greek political spectrum have come to recognize that the nationalistic course Greece pursued from 1991 through 1994 was totally counterproductive and resulted in its isolation in the Balkans, as well as in Europe more generally. As a result, Greece has begun to pursue a more pragmatic policy designed to end its diplomatic isolation.

This more moderate, pragmatic policy has been reflected in particular in Greek policy in the Balkans. Relations with Albania, which had deteriorated after the arrest of five members of the Greek minority in the fall of 1994, have improved visibly since March 1995. The dispute with Macedonia has been partially defused as a result of the Interim Agreement signed in September 1995. Relations with Bulgaria and Romania have also been strengthened.

In addition, Greece has begun to play a more active role in promoting regional cooperation. At the end of October 1995, Greece proposed the formation of a Regional Council of Balkan states, composed of Albania, Bulgaria, Romania, Greece, and the countries of the former Yugoslavia. While the main purpose of the council is to promote cooperation in the fields of economics, technology, the environment, culture, and tourism, Greek officials have said cooperation could eventually be extended to security issues as well.

The shift to a more moderate, pragmatic foreign policy, moreover, has coincided with important structural shifts in Greek domestic politics. The collapse of communism in Eastern Europe has discredited the Greek left and has reduced the saliency of anti-Americanism, once the ideological mainstay of the Greek left, including PASOK.[4] PASOK has undergone a significant evolution in recent years, both in internal and external policy, and has increasingly become more of a traditional European social democratic party. In foreign policy, it no longer questions Greece's membership in NATO or the EU. In the domestic area, it has moved away from the expansionist economic and social welfare policies of the early 1980s and has adopted a "Thatcherite" austerity program designed to address Greece's structural economic problems, especially its high public deficit.

As a result, the differences between the New Democracy—the main conservative opposition party—and PASOK have significantly diminished. The main division in Greece today is not between the New Democracy and PASOK but between the pro-European "modernizers," who advocate major structural reforms to enable Greece to integrate more fully into Europe, and the more traditional-minded nationalists and populists, who advocate slower structural change. These divisions cut across lines and have tended to blur the distinctions between the two major parties. Indeed, on key foreign policy issues, there is a general consensus between the two parties.

---

[4]PASOK stands for the Pan-Hellenic Socialist Movement.

PASOK, however, is entering a period of important change. The Papandreou era is rapidly coming to a close. Papandreou's departure from the political scene will have an important impact on the party and could accelerate the transformation of PASOK into a more open, modern, social democratic party, especially if the pro-European wing of the party succeeds in gaining the upper hand in the succession struggle.

Thus, on balance, Greece seems headed in the right direction. Its most serious problem remains the state of its economy, as well as the continued tensions with Turkey. There has been little serious progress in resolving differences with Turkey. However, if the Bosnian peace accord holds—a big if—Cyprus could become the next item on the Balkan agenda.

The intercommunal talks on Cyprus remain stalled. However, the signing of the EU-Turkish Customs Union agreement in March 1995 and the EU decision to open discussions with Cyprus about EU membership in 1997 have added a potentially new dynamic to the Cyprus question. Many Turkish Cypriots fear that, if a settlement is not reached soon, the EU may open discussions with the Greek part of the island, leaving the Turkish-Cypriots out in the cold. In the medium term, this may give the Turkish-Cypriots an incentive to be more flexible in the intercommunal talks on the island's future.

Ultimately, however, the keys to any settlement of the Cyprus issue lie in Ankara. Yet Turkey also has some incentive to see a settlement. Turkish-Cyprus is a drain on the Turkish economy—not an intolerable one, but a drain nonetheless. Moreover it is a burden in Turkey's relations within the EU and the United States at a time when Ankara is seeking to strengthen ties to both. Thus, in principle, Turkey has reasons to want to see the issue resolved. The problem is to find the right mixture of incentives and trade-offs that will induce the political leadership in Ankara to take the political risks necessary to obtain a settlement.

TURKEY: A NEW REGIONAL POWER?

Turkey has also emerged as a more important actor on the Balkan scene. Turkey has strong historical interests in the Balkans. For centuries, the Ottoman empire was the dominant power in the region. The collapse of the Ottoman empire prompted a Turkish retreat from the Balkans. However, the end of the Cold War and the outbreak of conflict in Bosnia have led to a revival of Turkish interest in the Balkans.

Turkey has sought to exploit the new opportunities opened up in the Balkans by the end of the Cold War to improve relations with a number of Balkan countries. The most important sign of this new activism has been the rapprochement with Bulgaria. For most of the Cold War period, relations between Bulgaria and Turkey were strained. However, since 1989, relations have undergone a significant improvement (Engelbrekt, 1991, pp. 9–10; Perry, 1992). This process was highlighted by the signing, in May 1992, of a Treaty of Friendship and Cooperation that calls for a broad expansion of ties in various political and economic areas.

Defense and military cooperation between Bulgaria and Turkey have also significantly increased over the last several years. In July 1991, a military delegation headed by Lieutenant General Radynu Minchev, Chief of the Bulgarian General Staff, visited Turkey—the first visit to Turkey by a Bulgarian chief of staff in the postwar period—and in March 1992, Bulgarian Defense Minister Dimitar Ludzhev became the first Bulgarian defense minister to visit Turkey in the postwar period.

In addition, in December 1991, the two countries signed a bilateral military agreement (the Sofia Document) designed to strengthen security and confidence along the Bulgarian-Turkish border. The agreement provides for each side to give the other advance notification of military activities within the zone of application and gives each side the right to conduct an inspection and two site visits on the territory of the other side beyond those provided for in the 1990 CFE accord. As part of this confidence-building effort, Turkey moved one battalion of ground forces and a tank battalion back from the Bulgarian-Turkish border in July 1992.[5]

---

[5]In November 1992, the Chiefs of the General Staffs of the two countries signed a second agreement on additional confidence-building measures—the Edirne Document. The Edirne Document supplements the Sofia Document signed in December 1991 and lowers the threshold for the reciprocal exchange of notification of and invitation to military maneuvers. It also provides for training and increased contacts between military representatives from the two countries.

Relations with Albania have also been strengthened, especially in the military sphere. In July 1992, the two countries signed an agreement on military cooperation. According to the terms of the agreement, Turkey will help modernize the Albanian army and help train Albanian officers (Zanga, 1993). While this military cooperation remains relatively modest, it has been viewed with considerable concern in Athens and has reinforced fears that a "Muslim arc" could emerge on Greece's northern border.

Turkey has strongly supported the Bosnian cause in the Balkans. Although the Bosnian Muslims are ethnic Slavs, not Turks, they are remnants of the Ottoman Empire. Turkey, therefore, feels a close cultural affinity to them and moral responsibility for their well-being. Moreover, there are over 2 million Bosnians in Turkey—the result of several waves of emigration since 1878, when the Ottomans began to withdraw from the Balkans. They represent a strong interest group and lobby in Turkey. Hence, Turkey cannot ignore developments in Bosnia, even if it wanted to.

The Bosnian conflict has also cast the "Muslim factor" in a new light. There are over 4 million Muslims in the former Yugoslavia. In addition, there are about 1 million Muslims, many of them ethnic Turks, in Bulgaria. Greece also has a large Muslim population (120,000), more than half of whom are ethnic Turks. To date, Turkey has consciously rejected playing "the Muslim card." However, pro-Islamic forces led by the Welfare Party (Refah) have made substantial gains in Turkish politics since 1994. If the Welfare Party were to come to power, the Muslim factor could become a more important issue in Turkey's foreign policy, especially in the Balkans.

THE RUSSIAN FACTOR

Russia has historically had a strong interest in the Balkans. However, since 1989, it has played only a marginal role there. Russia did have a brief moment of glory in February 1994 when it brokered a cease-fire with the Bosnian Serbs, but since then, it has largely been shunted to the sidelines. Russia played virtually no role in the final peace settlement in Bosnia.

Indeed, Russia's concern with Bosnia has largely been a function of other issues, rather than a reflection of close historical and religious solidarity with the Serbs. Moscow's primary concern has been to prevent a strengthening of NATO and to block NATO from becoming the cornerstone of a new security order in Europe. Russia took offense at NATO's growing involvement in the Bosnian crisis largely because it underscored Moscow's own impotence and strengthened the chances that NATO would become the key security organization in post–Cold War Europe.

The Bosnian issue also played an important role in Russian domestic politics. For many Russian nationalists, it was a convenient means of attacking Yeltsin and Foreign Minister Kozyrev. Bosnia rankled all the more because it underscored Moscow's impotence and lack of foreign policy influence in a region in which Moscow had traditionally had strong interests and had been an important player. In short, the internal debate over Bosnia has been part of the larger Russian debate about Russia's "national interests" that has raged for the last several years and has preoccupied so many members of the Russian elite.

However, despite Russia's strong historical interest in the region, the Balkans are unlikely to become a major focal point of Russian policy. For the next decade, Russia will be preoccupied with its own internal problems and with developments in the near abroad. This will leave little time or energy for Balkan affairs. Moreover, the real issue in the post-Dayton era in the Balkans is likely to be economic reconstruction. Here, Russia has little to offer. The key player is likely to be the EU, not Russia.

This does not mean that Russia is likely to forswear all interest in the Balkans. Bulgaria could again emerge as an important focal point of Russian attention. The election of a postcommunist government in Bulgaria in December 1994 has opened up new opportunities for closer cooperation between Moscow and Sofia, especially in the economic realm. As noted earlier, the BSP government has also adopted a more equivocal attitude toward NATO membership—a fact which has not gone unnoticed in Moscow. However, as long as President Zhelev—a strong supporter of NATO—remains in power, Bulgaria is likely to remain on a pro-Western course. The real issue is what happens after he leaves.

Russia's relations with Greece have improved over the last several years. But the prospects for any significant rapprochement remain limited. Greece's main concern is to repair relations with its Balkan neighbors and the EU. Its future lies with a stronger and more prosperous Europe. Some Greeks see Russia as a possible counterweight to Turkey in the Balkans. But on key issues of concern to Greece—especially Cyprus and the Aegean—the United States and the EU are the critical actors, not Russia.

Russia's relations with Turkey, on the other hand, seem likely to become more strained. Turkey's expanding ties to the Muslim countries in Central Asia and the Caucasus (especially Azerbaijan) have made Russia very nervous and have sparked a new, albeit muted struggle for influence in the region between Russia, Iran, and Turkey (Fuller, 1992). Old historic geopolitical rivalries are beginning to reemerge. Indeed, to a certain extent, there may be a replay of the "Great Game" in a new geopolitical context.

Energy issues, above all the struggle for control over Caspian oil, are likely to give this geopolitical rivalry greater impetus. Whoever controls these resources—and access to them—is likely to be the dominant political power in the region. The struggle over the routing of the pipeline to carry Caspian Sea oil is more than a struggle over economies. At its heart, it is a geopolitical struggle over control of the Central Asian–Caucasian geopolitical space.

How this geopolitical rivalry plays itself out over the next decade will have enormous implications—not only for the future of Central Asia and the Caucasus, but also for Western policy toward Russia, Turkey, the Middle East, and the Persian Gulf. To date, Western policymakers have tended to view these areas separately and largely in isolation. The emerging geopolitical rivalry between Russia, Iran, and Turkey in Central Asia and the Caucasus, however, suggests that these issues are becoming more closely linked and that policy toward one area is likely to have increasing impact on policy toward the other.

THE WESTERN ROLE

Despite the Dayton agreement, the prospects for peace and stability in the Balkans remain tenuous and uncertain. As noted earlier in this chapter, several important flash points continue to exist that could create future problems, especially in Kosovo and Macedonia. Thus, if stability is to be achieved in the region, these problems will need to be addressed and be part of a larger Balkan peace process.

U.S. political engagement will be critical. The issue for the United States is not what happens in Bosnia or Kosovo, per se, but the larger impact that developments there can have on regional stability and Alliance relations. As the Bosnian conflict has shown, it is impossible to seal off developments in the Balkans hermetically from the larger issues of European security, particularly NATO's future. If things go badly in Bosnia or elsewhere in the Balkans, they will have an impact on Alliance interests and relations. NATO enlargement, in particular, could become more difficult, perhaps even derailed. Thus, the United States cannot afford to simply "leave the Balkans to the Europeans." It needs to work actively with its European allies to help shape a stable security framework in the region.

In particular, the United States should encourage closer economic and regional cooperation among the Balkan countries. Without sustained economic growth and development, many of the fledgling democratic reforms in the region are likely to falter, plunging the countries in the region into a new round of instability and ethnic violence. The Greek initiative to form a Regional Council of Balkan States is a potentially important development in this regard. But more needs to be done on the multilateral level as well, particularly by the European Union.

The EU is likely to be the leading actor in the next phase of Balkan politics, which will focus strongly on the economic reconstruction of the area. This reconstruction effort should not be limited to Bosnia: A comprehensive strategy for the economic reconstruction of the Balkan region as a whole will be necessary. This should include the conclusion of a common set of economic and trade agreements with the countries of the region (except Bulgaria and Romania, which already have association agreements, and Slovenia, which is a potential candidate for early membership in the EU). This would provide an important framework for developing relations with the entire southern Balkans and enhance the prospects for the emergence of greater regional stability over the long run.

REFERENCES

"Bukarest bemüht sich um Image-Korreckur," *Neue Zürcher Zeitung,* October 30, 1995.

Engelbrekt, Kjell, "Relations with Turkey: A Review of Post-Zhivkov Developments," *Report on Eastern Europe,* April 26, 1991, pp. 9–10.

Fuller, Graham E., *Turkey Faces East: New Orientations Toward the Middle East and the Old Soviet Union,* Santa Monica, Calif.: RAND, R-4232-AF/A, 1992.

Glenny, Misha, "Macedonian Birth Pangs," *New York Review of Books,* November 16, 1995, pp. 24–28.

Larrabee, F. Stephen, *The Volatile Powderkeg: Balkan Security After the Cold War,* Washington, D.C.: American University Press, 1994.

Perry, Duncan, "New Directions for Bulgarian-Turkish Relations," *RFE/RL Research Report,* October 16, 1992, pp. 33–39.

_____, "On the Road to Stability—or Destruction," *Transition,* August 25, 1995, pp. 40–48.

Rüb, Mattias, "Historische Versöhnung nach deutsch-französischen Vorbild," *Frankfurter Algemeine Zeitung,* October 12, 1995.

Zanga, Louis, "Albania and Turkey Forge Closer Ties," *RFE/RL Research Report,* March 12, 1993, pp. 30–33.

Chapter Five

EAST CENTRAL EUROPE

F. Stephen Larrabee

Since the collapse of communism, the countries of East Central Europe have made substantial progress toward the establishment of stable democratic systems and the creation of market economics.[1] The three "fast track" countries—Hungary, Poland, and the Czech Republic—have emerged from the recession that followed the initial efforts to move toward market reform in 1990 and 1991 (Dove and Robinson, 1995).[2] Throughout the region, growth rates are rising, while inflation has dropped significantly. Privatization has also taken root. Today, between 60 and 70 percent of the region's economies are in private hands. Trade has been reoriented toward the industrial countries of the West. The countries of the region now conduct over half their trade with the European Union (EU). Foreign direct investment has also risen steadily.

THE POSTCOMMUNIST RESURGENCE

The transitions in East Central Europe, however, are far from complete or irreversible. Indeed, East Central Europe has begun to witness a certain sense of "reform fatigue" lately. Over the last two years, postcommunist parties have made a comeback throughout

---

[1]For the purposes of this chapter, the term East Central Europe refers to Hungary, Poland, the Czech Republic, and Slovakia. A separate chapter deals with the Balkans.

[2]See also "Central Europe: The Winners" (1995).

Eastern Europe, regaining power in Lithuania, Poland, Hungary, and Bulgaria.[3] This has caused concern in some Western capitals.

The recent shift to the left in East Central Europe, however, needs to be seen in perspective. It does not reflect a longing for a return to communism as much as it does a disenchantment with the economic and social impact of reform. Many East Central Europeans had gotten used to what Wojciech Gebicki and Anna Marta Gebicka have aptly termed the "nanny state" in which the individual could, to a large extent, dispense with personal responsibility and leave everything to "the system" (Gebicki and Gebicka, 1995). Free education and health care, as well as life-long security, were taken for granted. This "culture of entitlement" was reinforced by exaggerated expectations of a rapid transition to economic prosperity, which was associated in the minds of the public in East Central Europe with Western-style democracy and a flourishing market economy. Thus, many East Europeans were psychologically ill-prepared for the harsh dislocations that accompanied the collapse of communism and the transition to a market economy.

The postcommunist parties were able to capitalize on this disenchantment and turn it to their political advantage, especially in Poland and Hungary. This does not mean, however, that East Central Europe is heading back toward communism. Far from it. The postcommunist parties in Poland and Hungary are strongly committed to the development of a market economy and Western integration. Indeed, in some cases, they have pushed these policies faster than their predecessors. In Hungary, there was considerable concern when Prime Minister Gyula Horn fired the liberal Minister of Finance Laszlo Bekasi and replaced him with Lajos Bokros. Many Western observers saw Bekasi's ouster as an indication that the Horn government was backtracking on reform. But Bokros has implemented an austerity program that has gone even further than Bekasi's program.

Similarly, in Poland, Finance Minister Gregorz Kolodko's economic program was essentially a continuation of the reform program carried out by the Suchocka government. The main center of opposi-

---

[3]Here, *postcommunist* refers to former communist parties that have regained power.

tion came from Solidarity, which objected to many of the program's tough austerity measures. This underscores the way in which old categories and ways of thinking no longer fit the new realities in East Central Europe.

Nor should the leaders of the postcommunist parties simply be dismissed as "communist retreads." Many, like Gyula Horn, the current Prime Minister of Hungary, had already begun to undergo a political evolution in the late 1970s and 1980s, well before the Berlin Wall fell. As foreign minister in the last communist government in Hungary, Horn played a key role in precipitating the Wall's eventual collapse: It was his decision in the Fall of 1989 to allow the transit of East German citizens to West Germany that set in motion the mass exodus that eventually precipitated the disintegration of the German Democratic Republic and the unification of Germany.

Generational factors also play a role in the postcommunist resurgence. Many of the younger postcommunist leaders were too young to be closely associated with the policies of the previous communist regimes. They have been able to shed their communist past and project themselves as young, forward-thinking technocrats more concerned about the future than the past, whereas their democratic opponents have often appeared to be still fighting the ghosts and battles of the past. Poland represents a case in point. In the Presidential election in November 1995, Alexander Kwasniewski, the leader of the Democratic Left Coalition (SDL) and a former minister in the Jaruzelski government, consciously sought to portray himself as a modern social democrat concerned with the future, whereas then-President Lech Walesa often seemed to have little to offer except his anticommunist credentials. Kwasniewski's victory was aptly foreshadowed in his reply to charges about his communist past during one of the televised debates with Walesa: Walesa, he said, had made important contributions to ending communism, but now it was "time to move on." A majority of Poles, especially the younger ones, appear to have agreed.

In short, the reasons for the success of postcommunist parties in East Central Europe have more to do with a desire to preserve certain aspects of the "nanny state" and cushion the shock of reform than with a desire to see the return of communism. The return of the postcommunists should be seen as a natural part of the evolution toward democracy. The problem of "reform fatigue" currently evident in East Central Europe is not unique. Other countries undergoing transitions from authoritarian to democratic rule have witnessed similar problems. Spain, for instance, faced a similar period of public disillusionment and disenchantment (*desencanto*) in the late 1970s as it sought to consolidate the democratic transition opened up by Franco's death (Maravall and Santamaria, 1986, pp. 93–94). Nevertheless, the postcommunist resurgence is an important reminder of how fragile the social consensus in East Central Europe still is and how far the transitions still have to go before they are fully consolidated.

The real danger in East Central Europe is not that the region will slide back into communism. Communism has been too discredited for that. Rather, the danger is the emergence of a "populist managed economy," similar to that in Argentina under Peron, that promises quick-fixes and relies on a combination of nationalism, "controlled" democracy, and state intervention.[4] Something akin to this model, in fact, appears to be emerging in Slovakia where Prime Minister Vladimir Meciar has slowed down the privatization program and tried to use nationalism to consolidate his power (see below).

ECONOMIC REFORM AND POLITICAL CONSOLIDATION: A BALANCE SHEET

Slovakia, however, is the exception in East Central Europe, not the rule. The other three East Central European countries have made important steps toward creating the foundations of stable democratic systems and market economies. The pace and modalities of reform, however, have varied from country to country—as has the degree of success.

The Czech Republic has led the pack. Under the leadership of Prime Minister Vaclav Klaus, the head of the right center Civic Democratic Party (ODS), the Czech Republic has made steady progress toward market reform and political stability. In 1995 the Czech economy witnessed a 4-percent growth rate. This is expected to rise to over 5

---

[4]For an insightful discussion of this point, see Gebicki and Gebicka (1995, pp. 137–138).

percent in 1996. Inflation has dropped to under 10 percent, and unemployment is about 3 percent—the lowest in Eastern Europe. The private-sector share of the Czech Republic's gross domestic product (GDP) is the highest in East Central Europe, nearly 80 percent. Prague has also introduced the most extensive large-scale privatization program in East Central Europe. In November 1995, the Czech Republic became the first postcommunist state in Eastern Europe to join the OECD.

The success of Klaus' reform program has prevented the emergence of the type of popular dissatisfaction that has propelled postcommunist parties back to power elsewhere in East Central Europe. However, many of the hardest tests for the Czech economy are still in the future. Unemployment is low, for instance, because the government has been reluctant to implement a tough bankruptcy law. This has helped perpetuate hidden unemployment. If the government pursues a tougher policy toward bankruptcies, unemployment—the lowest in East Central Europe—could grow. Similarly, the privatization program has been less successful than appears at first glance. Big banks, many of them state owned, hold a large portion of the privatized equity. Czech industry also remains highly inefficient. Thus, the Czech "economic miracle" may lose some of its luster once the full impact of the reforms begins to hit and as the Czech economy becomes more integrated into the world market.

Moreover, the political stability that has characterized Czech politics since 1991 has recently begun to erode. While Klaus remains the most popular politician in the Czech Republic, his coalition, dominated by ODS, has been plagued by increasing scandals and internal bickering lately. One of his coalition partners, the Christian Democratic Party, has split, and there is growing disenchantment with Klaus' autocratic style, as well as with his tough stance on social spending. The Social Democrats, the main opposition party, have also increased their strength. Klaus' party, the ODS, still seems likely to gain the most votes in the election in June 1996. However, the recent strains within the ruling coalition suggest that Klaus' domination of Czech politics can no longer be taken for granted.

After a difficult start, Poland's pursuit of a policy of shock therapy has begun to show results. In 1994, the Polish economy witnessed a 5-percent growth rate—the highest in Eastern Europe—and it is expected to be 6 percent in 1995. This growth was led by an impressive increase in investment and exports rather than domestic consumption. Privatization has also expanded significantly. More than 60 percent of the Polish work force is now employed in the private sector and produces more than 65 percent of the country's GDP. This turnaround has begun to attract increased foreign investment. Indeed, many economists believe that Poland, not the Czech Republic, will be the real economic tiger in the region over the long run.

Poland's political transition, however, has been less smooth. The ruling coalition between the Peasant Party and the Democratic Left Coalition (SDL) has been wracked by infighting and feuds, hindering the development of a coherent economic and social policy. These problems were compounded by former President Lech Walesa's continued attempts to systematically weaken the ruling coalition and increase his own political stature and electoral prospects. These tactics resulted in a continuous battle with the government and parliament and inhibited the implementation of a coherent reform program. At the same time, Walesa's persistent efforts to weaken the minister of defense and bring the military under the direct control of the president undercut previous efforts to establish civilian control of the military and damaged Poland's image in the West. Indeed, by the end of his tenure, many Poles—including many former colleagues in Solidarity—had come to view Walesa as more of a threat to democracy than its protector.

These developments were responsible for a significant drop in Walesa's popularity and his eventual loss of the presidential election to SLD leader Alexander Kwasniewski. However, as noted earlier, Kwasniewski's election should not be seen as a signal that Poland is about to return to Soviet-style communism. Kwasniewski campaigned as a Western-style social democrat and as the man most capable of leading Poland into a new, modern era. During the campaign, he pledged to continue to press for Poland's integration into NATO and the EU. His election is thus not likely to result in any pronounced shift in Poland's basic foreign policy orientation. Indeed, Kwasniewski may push closer ties to NATO and the EU even more forcefully than his predecessors, to dispel any doubts in the West about his pro-Western orientation.

On the domestic side, Kwasniewski's election may have important positive benefits. First, and most important, it is likely to end the divisive power struggle between the president and the government that marked Walesa's tenure as president and also facilitate a smoother working relationship between the president and the parliament. Unlike Walesa, Kwasniewski is not likely to engage in a constant effort to block legislation and destabilize the government. Aware of widespread concerns that all three branches of the government—the parliament, executive, and presidency—are in the hands of former communists, Kwasniewski is likely to act as an above-party president rather than "president of the SLD."

Civil-military relations are also likely to improve. Unlike Walesa, Kwasniewski supports the subordination of the General Staff to the minister of defense rather than the president. In addition, he will probably replace some of the top officers on the General Staff who favored greater autonomy for the military, particularly General Tadeusz Wilecki, the chief of the General Staff. These moves should end the destructive infighting and bickering over who controls the military that poisoned civil-military relations under Walesa. They will also be welcomed in the West and enhance Poland's chances to obtain NATO membership.

Walesa's defeat raises the question of what role he will play in Polish politics in the future. At 52, he is too young to simply "retire." Moreover, temperamentally he is ill suited to play the role of elder statesman. He found it difficult to sit in "splendid isolation" in Gdansk and let Tadeusz Mazowiecki, Poland's first noncommunist prime minister, run the government in 1989 and 1990. And he is likely to find it even more difficult to gracefully fade into the political sunset now. He may, therefore, continue his struggle on the outside, either by founding his own party and/or trying to unite the fractious anticommunist Polish right.

Hungary's transition has been somewhat uneven. On the political side, the transition has been relatively smooth. The foundations of a stable political democracy have been established, and the extreme right—which a few years ago seemed like it might pose a political threat—has been marginalized. A broad consensus on foreign policy—especially on Hungary's membership in the EU and NATO—exists. Relations with Hungary's neighbors have also improved.

Hungary's economic transition, however, has been more problematic. The Hungarian Democratic Forum (MDF), led by former Prime Minister Jozsef Antall, adopted a more gradual approach to reform than Poland or the Czech Republic. Nevertheless, the reform program created severe dislocations. Hungary witnessed a sharp drop in living standards, growing inequality of incomes, and a sharp rise in public dissatisfaction with reform.[5]

This growing disenchantment with reform eroded support for the MDF and enabled the Hungarian Socialist Party (HSP), led by Gyula Horn, the foreign minister in Hungary's last communist government, to win an overwhelming victory in the May 1994 elections. The return of the Socialists prompted concern that the Horn government might slow down economic reform. However, after some initial hesitation, the government launched a tough austerity program in March 1995. These austerity measures have contributed to a revival of the Hungarian economy.[6] The budget and current-account deficits have been reduced, and inflation has dropped slightly. However, Hungary's high indebtedness remains a problem, as does the high percentage of GDP spent on welfare.

Foreign policy has been marked by continuity as well. Like its predecessors, the Horn government has given top priority to Western integration, especially membership in the EU and NATO. The one real change in the foreign policy field has been the Horn government's approach to the Hungarian minority issue. In contrast to the MDF, the Horn government has down-played the minority issue and given a higher priority to trying to reduce tensions with Slovakia and Romania. A bilateral treaty was signed with Slovakia in March 1995. However, relations continue to be marred by differences over the treatment of the Hungarian minority.[7] Relations with Romania have also gradually improved. The two sides hope to be able to conclude a long-delayed bilateral treaty in early 1996.

---

[5]For details, see Andorka (1994).

[6]See "Hungary Revived by Tough Medicine" (1995).

[7]The most recent differences have occurred over the draft of a language law that makes Slovak the only official state language. Hungary claims the law violates the spirit, if not the letter, of the Hungarian-Slovak bilateral treaty signed in March 1994. See "Budapests schwieriges Verhältnis zu Bratislava" (1995).

Slovakia has made the least progress of all the four East Central European countries toward the creation of a stable democratic system and market economy. In fact, the pace of economic and political reform has slowed since Prime Minister Vladimir Meciar's return to power in September 1994 (Boland, 1995). The mass privatization program initiated earlier has been cut back and the role of the state in the Slovak economy has increased. Meciar has also sought to curb the independence of the media and engaged in an open effort to oust President Michal Kovac. These developments have damaged Slovakia's chances for early membership in the EU and NATO.[8]

MILITARY REFORM AND DEFENSE POLICY

Since the collapse of communism, the countries of East Central Europe have carried out an extensive process of military reform. The extent and depth of this process has varied from country to country.[9] It has been the most extensive in the Czech Republic, whereas in Slovakia it is still in its embryonic stage. All the countries of East Central Europe, however, face severe budgetary constraints that inhibit their ability to carry out reforms.

Civilian control of the military also remains a problem. Here again the pace and extent of reform has varied. It has gone the furthest in the Czech Republic, which carried out an extensive review of the officer corps and has put civilians in many of the top posts in the Ministry of Defense. It has been most problematic in Poland, largely due to President Walesa's effort to weaken the position of the minister of defense and subordinate the military directly to the president (see below). Hungary made some initial progress, but has shown some signs of backsliding since the return of the Socialist Party to power in May 1994. In Slovakia, civilian control barely exists. Moreover, the minister of defense is a member of the ultranationalist Slovak Nationalist Party.

---

[8]In October 1995, the EU issued an official warning to Slovakia, emphasizing the need to respect the norms of democratic pluralism. The United States also expressed a similar warning. See "Europäische Union mahnt Slovakei" (1995).

[9]For a detailed discussion, see Szayna and Larrabee (1994).

In all the countries, however, civilian control of the military remains superficial and often does not extend much below the deputy minister or state secretary level. Most civilian appointees do not have deep or detailed knowledge of defense issues, and they remain dependent on the military for advice and analysis. There is no "counterelite" or cadre of civilian specialists who can challenge the military's views and provide an alternative viewpoint, such as exists in the United States and many countries of Western Europe. This has inhibited the establishment of effective civilian control over the military.

All the East Central European countries have reduced and restructured their militaries. The Czech Republic has probably made the most progress. The former Czechoslovak army (both personnel and equipment) was divided on a 2-1 basis when the Czech Republic and Slovakia became separate states on January 1, 1993. The new Czech army has been radically reduced and restructured. It has been cut from a force of 106,477 men at the time of the split on January 1, 1993, to about 65,000 by 1995—a 40-percent reduction. About half of these are expected to be professionals.

The ground troops, which numbered about 43,000 men in June 1993, will be cut by one-third, to about 28,000 men, by 1995. The old system based on divisions has been replaced by a brigade-based system. The new structure consists of the following:

- The Territorial Defense Force of 15 brigades, with each brigade operated in peacetime by only a skeletal garrison

- The Expeditionary Force, composed of seven mechanized brigades (four in Bohemia and three in Moravia)

- The Rapid Deployment Force, made up of one brigade of 3,000 men, which is specifically configured to be able to mesh with NATO units.

The military restructuring process, however, has had to be implemented in an atmosphere of economic austerity, which has put severe constraints on resources that could be devoted to defense. Financial constraints, for instance, initially led the Czech Ministry of Defense to decide to modernize its fleet of MiG-21s rather than buy used F-16s, as Poland is considering doing. However, the decision was so strongly criticized by many Czech parliamentarians and experts that the ministry later suspended it.

Financial concerns also appear to be inducing the Czechs to take a more positive attitude toward defense cooperation with their Visegrad neighbors. Poland and the Czech Republic are currently discussing ways to gain economies of scale through pooled arms production and weapon imports, as well as swaps of military equipment (Tigner, 1995; McNally, 1995). One idea currently under discussion involves Poland supplying Sokol attack helicopters to the Czech Republic in return for Czech L-159 trainer jets.

Poland's military also has undergone a significant restructuring. A new draft doctrine was announced in July 1992. The new doctrine calls for creating a relatively small military force—about 200,000 troops (0.5 percent of the population)—and Territorial Defense units. The doctrine also reiterates Poland's determination to seek membership in NATO and the WEU (National Security Bureau, 1992, p. 13).

In connection with its reoriented defense doctrine, Poland has expanded the number of military districts from three to four (two in the east and two in the west) and has begun to redeploy its forces, stationing more troops in the east. However, this process is costly because a new infrastructure must be built, and Poland faces a severe budgetary crunch. Thus, Poland can only implement these changes gradually. In addition, the new doctrine emphasizes creating lighter, more-mobile forces that can react quickly to local conflicts, especially with Poland's immediate neighbors.

In carrying out its reform program, Poland has placed a high priority on upgrading its equipment and making it compatible with NATO equipment, as well as on preparing for joint peacekeeping missions with NATO. A first battalion is expected to be ready by the end of 1995. The battalion, according to Polish officials, will be fully compatible with NATO command and control operational and equipment standards.[10]

Upgrading its air defense system and modernizing its air force are also top Polish priorities. As part of a five-year modernization plan,

---

[10]See the interview with Polish Defense Minister Zbigniew Okonski (1995).

Poland is currently shopping around for replacements for 200 MiG-21s, which are obsolete. Like the Czech Republic, Poland is also considering buying used F-16s. Moreover, as noted earlier, Poland is investigating possible ways to gain economies of scale through pooled arms production, as well as through swapping military equipment with other East Central European countries, particularly the Czech Republic.

The military reform and modernization process in Poland has been inhibited, however, by strong economic constraints. Polish defense spending has declined by 38 percent in real terms since 1986. As a percentage of GDP, it fell from 3.2 percent in 1986 to 1.9 percent in 1993.[11] As a result, little money has been allocated for procuring new weapons.

Perhaps the biggest obstacle to reform, however, was the constant infighting between President Walesa and a succession of defense ministers over control of the armed forces.[12] This infighting not only weakened effective control over the military but also inhibited the implementation of a coherent reform program. However, as noted earlier, these problems should significantly diminish now that Walesa is no longer president. The new constitution, which is currently being drafted, is expected to clarify the powers of the president and prime minister in the area of defense and is likely to subordinate the military directly to the Ministry of Defense.

Hungary adopted a new defense doctrine in March 1993. The new doctrine—or defense principles—attempts to bring Hungary's military force structure and tasks into harmony with Hungarian national interests. It emphasizes that Hungary has no "main enemy" and that the task of Hungary's military forces is solely defensive. The doctrine identifies small-scale incursions, provocations, and violations of

---

[11]Figures provided by the Polish Ministry of Defense, March 1993.

[12]The most notorious example of Walesa's backstage maneuvering and intrigues to increase his control over the military was the so-called "Drawsko affair" in October 1994, in which Walesa reportedly met privately with a group of senior Polish officers and encouraged them to speak out against Defense Minister Piotr Kolodziejczyk. Kolodziejczyk, a retired military officer and former ally of Walesa's, was fired shortly thereafter, largely because he resisted Walesa's efforts to subordinate the General Staff to the president rather than the defense minister. For the background, see "Walesa fordert Verteidigungsminister zum Rücktritt auf" (1994).

Hungary's airspace as the most likely threats. In case of general war or large-scale aggression against Hungary, the principles leave open the possibility of outside assistance by friendly states.

To handle possible small-scale excursions by neighboring states, Hungary is developing a rapid-reaction force. The border guards under the Ministry of the Interior are also being augmented, and Hungary has established a special center for training peacekeeping forces. The army has also been reorganized into three regional commands. Hungarian forces, deployed primarily in the west under the old Warsaw Pact system, will now be distributed more or less equally over Hungarian territory, with highest-readiness forces in areas where Hungary is most likely to face attack (south). The army has also been reduced from a high of 150,000 troops in 1989 to about 90,000 at the end of 1995. It is expected to be further reduced to 60,000 by 1998 and to 50,000 by 2005 (Gorka, 1995).

Hungary has also launched an ambitious ten-year program to revamp its antiquated air-defense network. The new program, approved by parliament in September 1995, envisages a two-phased program. The first phase will begin in 1995 and will be completed in the year 2000. The second phase, beginning in 2000, will be completed by 2005. The program is designed to reinforce early warning capabilities along Hungary's southern border and prepare the country for a contributing role in NATO.

In addition, the Hungarian army will witness an important changing of the guard at the top in the near future. A number of high-ranking officers, including Army Commander Janos Deak and Chief of Staff Sandor Nemeth, are scheduled to retire in 1996, opening up key positions for younger, Western-trained officers. As part of a reorganization of the military approved by parliament in June 1995, the position of Army Commander and Chief of the General Staff will be merged. The new Chief of the General Staff is likely to be selected from among candidates who have studied at a U.S. or West European military academy.

Slovakia is just beginning to create its own armed forces and to develop its own military concept. The new Slovak army is expected to comprise about 35,000 men—less than half the size of the current Hungarian army. As a part of the division of military assets when the

Czechoslovak Federation split, Slovakia received ten MiG-29s, and it acquired five more from Russia at the end of 1993 as an offset against Russia's debt to Slovakia. However, military reform has proceeded slowly, both for political and economic reasons. The fact that the Defense Ministry is controlled by the ultranationalist Slovak National Party has hindered the establishment of civilian control over the military. Popular support for NATO membership is also weaker than in other East Central European countries.

REGIONAL COOPERATION

In the initial period after the collapse of communism, there was a marked growth and interest in regional cooperation within East Central Europe. The most prominent example was the case of the Visegrad group, which was formally established in February 1991. Originally composed of Hungary, Poland, and Czechoslovakia, the group was expanded to include the Czech Republic and Slovakia after the two republics became independent states on January 1, 1993. Cooperation within the group was largely ad hoc and informal. Initially, it was designed to coordinate an approach to Western institutions, especially the EU.

Over the last several years, however, cooperation within the Visegrad group has stagnated, largely due to resistance and opposition on the part of the Czech Republic. Czech Prime Minister Vaclav Klaus has consistently opposed any effort to "institutionalize" Visegrad cooperation, fearing this will inhibit the Czech Republic's integration into the EU. The Czech go-it-alone strategy has hindered the development of any far-reaching cooperation within the group.

However, some deepening of economic cooperation has occurred. In 1992, the four countries agreed to set up a free-trade zone (CEFTA). This has helped to promote closer trade and economic relations within the region. Some limited cooperation has also occurred in the defense field. As noted above, Poland and the Czech Republic have recently begun to discuss some pooling of arms production and swapping of military equipment—a move which may indicate that the Czech Republic is beginning to rethink its previous go-it-alone strategy.

In October 1995, Slovenia became a full member of CEFTA. The inclusion of Slovenia could give new impetus to cooperation within the group, especially in the economic area. This move reflects Slovenia's effort to distance itself from its Balkan neighbors and strengthen its ties to Europe. Slovenia has made significant progress toward the creation of a strong market economy and a stable democratic system, since its departure from the former Yugoslav federation in June 1991. Moreover, it has more in common culturally, economically, and politically with the Visegrad countries than with the other members of the former Yugoslavia or countries in the Balkans. (It has the highest per capita income in all of Eastern or Central Europe.) Thus, its inclusion in CEFTA makes political and economic sense. Membership in CEFTA also enhances Slovenia's chances of obtaining membership in the EU and NATO down the line.

Cooperation within the Central European Initiative (CEI), another important regional organization, has also languished. Originally formed in 1978 to coordinate cooperation between the border regions of Yugoslavia, Italy, Germany (Bavaria), Austria, and later Hungary, the CEI (initially called the Alpen-Adria Cooperation) was expanded in May 1990 to include Czechoslovakia. A year later, Poland joined. In July 1992, Croatia, Slovenia, and Bosnia-Herzegovina were added, and in July 1993, Macedonia became a member.

The conflict in the former Yugoslavia, however, has seriously retarded cooperation. In addition, the political crisis in Italy—one of the original promoters of the group—has reduced Italy's interest in actively using the group to promote greater regional cooperation. Finally, cooperation has been hindered by differences over minority issues, especially between Hungary and Slovakia. As a result, regional cooperation within the CEI has lost considerable momentum of late.

Recently, efforts have been made to infuse new life into the group. At its meeting in Warsaw in October 1995, the leaders of the CEI decided to extend a formal invitation to Ukraine, Belarus, Romania, Bulgaria, and Albania to join the group at its next meeting and pledged to undertake efforts to assist in the reconstruction of Bosnia and Croatia.[13] They also agreed to set up a new information and documentation center in Trieste, to be financed by Italy.

The CEI provides a useful means for tying countries like Ukraine, Romania, Bulgaria, and Albania more closely to European and regional structures. But its influence is likely to remain modest. The group has little money of its own and is almost entirely dependent on funding by outside sources such as the World Bank and the European Bank for Reconstruction and Development. Moreover, the Czech Republic remains opposed to any effort to institutionalize cooperation, an attitude that is likely to inhibit any significant deepening of cooperation.

THE RUSSIAN FACTOR

Russian policy toward East Central Europe has shifted visibly since 1991. In the initial period after the breakup of the former Soviet Union, Russia was preoccupied with the problems of internal consolidation and essentially adopted a policy of "benign neglect" toward Eastern Europe. Indeed, Russia initially seemed to have no coherent policy toward Eastern Europe at all. Yeltsin's policy largely consisted of ad hoc initiatives designed to settle outstanding issues left over from the Soviet period, particularly resolution of debt questions. There was little effort to develop an overarching policy toward the region as a whole.[14]

Over the last two years, however, Russia has begun to define its interests toward Eastern Europe more clearly. This has manifested itself above all in Russia's effort to block East Central European membership in NATO. During his visit to Warsaw in August 1993, President Yeltsin implied that Russia would be willing to accept Polish membership in NATO. A few weeks later, however, apparently under strong internal pressure, he reversed himself. Since then Moscow has consistently opposed East Central European membership in NATO. In effect, Moscow appears to want to keep East Central

---

[13]See "Die Länder Mitteleuropas wollen Aufbau Bosniens und Kroatiens unterstützen," *Frankfurter Allgemeine Zeitung,* October 9, 1995.

[14]For a fuller discussion, see Larrabee (1993, pp. 153–164).

Europe as a neutral buffer zone, or a least deny its political-military potential to the West as long as possible.

This effort to block NATO expansion is seen by many East Central Europeans, especially the Poles, as proof that Moscow has not entirely given up its desire to retain some residual influence over East Central Europe's security options. Moreover, East Central European elites have been worried by Russia's more assertive policy toward the Baltic countries and "near abroad" lately. While they recognize that Russia is too weak at the moment to pose a real threat to their security, they fear that Western hesitation and vacillation over NATO enlargement could encourage Moscow to exert greater political pressure on Eastern Europe. At the same time, they see membership in NATO as an important "insurance policy" against the emergence of a resurgent Russia.

This is particularly true in the case of Poland. Polish leaders see Russia's main goal as the creation of a belt of economically and politically weak states in the region until such time as Russia is strong enough to redraw the spheres of influence in the region (Ananicz et al., 1995, pp. 11–12). They are concerned that further vacillation and hesitation by the West regarding NATO enlargement will reinforce instability in the region and encourage Moscow to step up its efforts to block enlargement. Alternatively, a stepped-up effort by Moscow to block enlargement could stimulate a strong anti-Russian backlash in Poland and throughout the region. In either event, the result would be a growth in regional instability and tension that would have negative repercussions for Poland's security and that of the other countries in the region.

For Poland, moreover, Kaliningrad (formerly Königsberg) poses a specific problem. After World War II, the Soviets turned Kaliningrad into a huge military base closed to all foreigners. The area remains highly militarized. Polish officials regard the large concentration of Russian troops, many of them in a high state of readiness, as a security problem. They would like to see the Kaliningrad district demilitarized or, failing that, see a significant reduction of the number and combat readiness of Russian troops.

However, this seems unlikely in the near future. Many of the troops in Kaliningrad are Russian troops that have been withdrawn from

Eastern Europe, Germany, and the Baltic areas. Given the acute housing shortage in Russia, Russia has nowhere to put these troops. Moreover, with the loss of air and naval facilities in the Baltic states, Kaliningrad's military significance has increased. Hence, the Russian military is likely to insist on maintaining a sizable military presence in the Kaliningrad area. Russian officials have even threatened to increase the level of troops and armaments there as a response to any enlargement by NATO, a move that would clearly heighten tensions with Poland.

UKRAINE: THE CRITICAL SWING FACTOR

Ukraine's future will also bear heavily on East Central Europe's security. Ukraine acts as an important geopolitical buffer between Russia and East Central Europe. The reintegration of Ukraine into Russia or a Russian-dominated Commonwealth of Independent States would have serious consequences for the security of East Central Europe, removing that geopolitical buffer and bringing Russian power back to East Central Europe's doorstep. Hence the countries of East Central Europe have a strong stake in the continued survival of an independent, stable, democratic Ukraine.

If Ukrainian independence were curtailed and Ukraine were reincorporated into a "Russian geographic space," both Poland and, to a lesser extent, Hungary and the Czech Republic would find their political room for maneuver constrained. Hence, they have made an improvement in relations with Ukraine a key element of their foreign policies and have sought to encourage Ukraine's integration into European institutions to the maximum extent possible.

Poland has made rapprochement with Ukraine the cornerstone of its Eastern policy. Poland was the first country to recognize Ukrainian independence (December 1, 1991), and bilateral relations between the two countries have developed rapidly since then. In May 1992, the two countries signed a Treaty of Friendship and Cooperation, which provides a comprehensive framework for future bilateral relations. In the treaty, both sides affirmed the inviolability of frontiers and renounced all territorial claims against each other. Military cooperation has also been strengthened.[15] In particular, the two countries have discussed the formation of a joint peacekeeping battalion.

However, Poland's ability to affect developments in Ukraine is limited. Given its current economic difficulties, Poland does not have the resources or capital to help Ukraine stabilize its economy in a serious manner. Moreover, Poland needs to be careful not to give the impression that it is seeking to build an anti-Russian alliance or axis with Kiev. Both these factors place objective limits on the degree of collaboration and cooperation that is likely to develop between Kiev and Warsaw in the future, especially in the security field.

So far, Ukraine has refused to join any CIS collective defense arrangements. But the pressure on it to do so could increase. Moscow has been pushing hard for closer economic and military cooperation within the CIS. In January 1995, Russia signed important agreements with Belarus and Kazakhstan that call for increased economic and security cooperation, including the creation of a customs union. And at the CIS summit in Almaty in February 1995, the members discussed setting up a joint air defense system.

Belarus' future orientation will also have an important impact on the security situation in East Central Europe. Belarus has traditionally been an invasion corridor to Poland. Hence, Poland has strongly supported Belarus' independence and encouraged Minsk's ties to Western political and economic structures. Bilateral cooperation has also been strengthened with Belarus. In June 1992, the two countries signed a Treaty of Friendship and Cooperation, which obliges both countries to respect existing borders and to renounce any territorial claims on the other. Military cooperation has also been expanded.

Polish officials have watched the growing economic and political rapprochement between Belarus and Russia since the Spring of 1993 with considerable concern. Closer defense cooperation between Russia and Belarus, especially the integration of Belarus into a seri-

---

[15]In February 1993, the two countries signed a military agreement that envisions an expansion of information exchanges and military training. The accord also calls for conducting joint exercises and developing joint activities in rear and technical supply of troops. However, both countries have emphasized that such cooperation does not constitute a security alliance and is not directed against other countries.

ous CIS collective defense arrangement, would have a negative impact on Poland's security. It would not only open up the possibility of the stationing of Russian troops on Poland's border, but could also increase the pressure on Ukraine to join a CIS collective-defense arrangement. However, Polish and East Central European leaders have little influence over Belarus' policy. Thus, there is little they can do to prevent the closer economic and political reintegration of Belarus with Russia.

TIES TO THE EUROPEAN UNION

Membership in the EU is a top priority for all the countries of East Central Europe. Hungary, Poland, and Slovakia have already formally applied for membership. The Czech Republic and Slovenia are expected to do so soon. All see EU membership as an indispensable part of their broader effort to "rejoin Europe" and integrate themselves into Western institutions.

The EU has also made Eastern enlargement an important priority. In December 1991, association agreements were signed with Hungary, Poland, and the former Czechoslovakia.[16] These agreements provided for a liberalization of trade and political consultations. Similar though slightly more restrictive association agreements were signed with Bulgaria and Romania at the end of 1992 and with the Baltic states this year.

At its meeting in Copenhagen in June 1993, the EU went a step further, inviting the East European countries (including Bulgaria and Romania) to become members of the Community as soon as they have met the economic and political requirements for membership. It also endorsed a package of trade concessions designed to speed up the reduction of tariffs and quotas blocking Eastern Europe's most competitive exports. At its summit in Essen (December 9–10, 1994), the EU laid down a "pre-accession strategy" based on modest trade liberalization and expanded political contacts. And in June 1994, the European Commission followed this up with a White Paper defining the *acquis communautaire* (the EU's body of laws and regulations)

---

[16]Separate association agreements were signed with the Czech Republic and Slovakia after the formal dissolution of Czechoslovakia on January 1, 1993.

that the East European countries must adopt to create a single market in labor, goods, and capital.

However, since then, doubts about the speed with which enlargement to the East can proceed have become more pronounced in Brussels. One reason is the magnitude of the adjustment that the East Central Europeans will have to make. The economic adjustments required of the new entrants dwarf those needed for earlier accessions, such as those of Spain and Portugal (1986) or Greece (1981). While there was a sizable gap in the standard of living between these entrants and those of the community members, the gap is much bigger in the case of the East Central European countries. This has raised serious doubts in Brussels about whether the East Central European countries could cope with the competitive pressures and high standards imposed by the EU's internal market.

The costs to the EU of adjustment are another factor. The EU Commission estimates that it would cost an additional 38 billion ecu ($50 billion) a year of current EU funds if regional aid policies were extended to the five countries that have already applied (Hungary, Poland, Slovakia, Latvia, and Romania) and the five that are likely to apply in the near future (the Czech Republic, Slovenia, Bulgaria, Estonia, and Lithuania). Extending the Common Agricultural Policy (CAP) eastward could be even more expensive. These considerations have caused some EU members to favor slowing down the process of enlargement (Barber, 1995).[17]

Moreover, little progress on enlargement is likely to occur until the EU sorts out its own institutional problems. These are supposed to be the subject of the EU Intergovernmental Conference (IGC), scheduled to begin in late 1996, which could run well into 1997. Until the EU makes key decisions about such issues as Economic and Monetary Union (EMU), a new budget, and institutional changes, such as qualified majority voting, it is unlikely to pay much attention to enlargement. Without such reforms, many EU members think managing a 25-member Union would be impossible. Thus, it could be the turn of the century before the EU begins to confront decisions about enlargement in earnest.

---

[17]Also see "The EU Goes Slow on Enlargement" (1995).

This has led some European officials to suggest new forms of transitional membership. However, this would challenge the traditional EU approach of insisting that new EU members adopt the *acquis communautaire* prior to membership. It would also encourage other member states to adopt a policy of "selective cooperation," favored by Britain, weakening the solidarity and cohesion of the Union. But with a Union of 25 members, such a policy may be inevitable anyway.

Ultimately, however, the issue of enlargement—and who is in the first tranche—will be decided not by the Commission, but by the EU heads of state. Here, broader political interests, not the Commission's narrower concerns, will be dominant. Moreover, as in the past, the process is likely to involve a considerable degree of political bargaining.

Germany's voice will be critical. Germany regards eastern enlargement as a top national priority: Bonn does not want to be the easternmost boundary of the EU. Moreover, German Chancellor Helmut Kohl has promised several East Central European leaders he will make sure their countries are in the EU by the year 2000—or thereabouts. Thus, he is unlikely to sign up to the EMU unless it sets the stage for enlargement. He can count on support from the Nordics (Finland, Sweden, and Denmark) as well as Austria—the so-called "German bloc"—and Britain, all of whom favor enlargement, albeit for different reasons.

In the end, these political considerations and imperatives are likely to drive the EU debate and count more than many of the Commission's narrower concerns. Moreover, some compromises may be found. New members, for instance, might not be granted access to the CAP and regional aid immediately, thus reducing the immediate financial burdens on the EU and providing the EU more time to restructure both programs. This would disappoint some new members, since access to EU cash is one reason for seeking membership. However, what the East Central European countries need most is not subsidies, but open markets.

NATO ENLARGEMENT

The uncertainties surrounding EU enlargement have lent even greater importance to the issue of NATO membership for the countries of East Central Europe. The desire for NATO membership is motivated by more than simply fear of a potentially resurgent Russia—although this is clearly an important factor, particularly in Poland. It reflects a more fundamental and deep-seated historical desire to be part of "the West." However, membership in the EU and the West European Union is an important—but not sufficient—means of achieving this goal, because the United States is not a member of these institutions.

One of the prime advantages of NATO in the eyes of the countries of East Central Europe is that it binds the United States to Europe and provides a means of maintaining a strong transatlantic connection. The leaders of the Visegrad countries are strong Atlanticists. They regard a strong U.S. political and military presence in Europe as an important stabilizing factor. As Czech President Vaclav Havel (1993) has noted:

> I am convinced that the American presence in Europe is still necessary. In the 20th century, it was not just Europe that paid the price for American isolationism; America itself paid a price. The less it committed itself at the beginning of European conflagrations, the greater sacrifices it had to make at the end of the conflicts.

For the countries of East Central Europe, membership in the EU and WEU is not enough because it does not provide *a strong transatlantic connection,* which they regard as a *sine qua non* for ensuring their security over the long run.

This helps to explain the initial skepticism and disappointment with which the Clinton administration's proposal for Partnership for Peace (PfP) was greeted in much of East Central Europe. Most of the countries, especially Poland, were initially cool to the idea because they saw it as a substitute for or an alternative to NATO membership. East Central European perceptions of PfP, however, have changed considerably since the Brussels summit. President Clinton's statement at the summit—repeated during his trip to Warsaw in July 1994—that the issue is "no longer whether but when and how NATO will expand" has reduced East European concerns about member-ship and created a much more positive attitude toward PfP on the part of the Visegrad countries. In contrast to the presummit period, PfP is now seen in East Central Europe as a *means to NATO member-ship, not a substitute for it.* As a result the East Central European countries now view PfP much more positively and have begun to use it as a vehicle for reshaping their defense policies in ways that will make it easier to achieve eventual membership.

Nevertheless, there is still a considerable way to go before NATO enlargement becomes a reality. The internal NATO study on enlargement, released in September 1995, lays out the general prin-ciples and modalities according to which enlargement should pro-ceed. But the Alliance is unlikely to "name names" until after the Russian presidential elections in June 1996—and perhaps not until early 1997.

Neither is it entirely clear who will be in the first tranche of new members. The most likely candidates are Poland, the Czech Republic, and Hungary. Slovakia's chances have receded lately, because of the slowdown in the pace of economic and political reform. Some NATO members are also pushing for including only Poland and the Czech Republic in the first tranche, leaving out Hungary. However, such a move would have a devastating political-psychological impact in Hungary, which played a key role in paving the way for the collapse of the Berlin Wall and the unification of Germany.

PROSPECTS FOR THE FUTURE

NATO enlargement is likely to emerge as *the* key European security issue facing the United States and its European allies over the next several years. How well it is managed will have a critical impact not only on stability in East Central Europe but on the evolution of the security order in Europe and the future of transatlantic relations more generally.

Several factors are likely to influence the enlargement debate. The first is Russia's attitude and policy. Russian opposition will compli-cate the process but is unlikely to stop it. There is a clear consensus within the Alliance against giving Russia a veto over Alliance decisions about its future. At the same time, the Alliance will need to find a way to work out a cooperation package with Russia and defuse its sense of isolation by integrating it more closely into all-European security structures.

A second factor will be the costs of enlargement. Yet, these need not be as large as many fear. The real cost driver is forward-stationed combat troops. However, unless there is a resurgent threat from Russia, there is no a priori need to station large numbers of combat troops in East Central Europe. The United States and its allies need to create the capability to reassure new members and reinforce them in times of crisis. The costs for such a posture should not be excessive, especially when spread out over ten years. Moreover, the costs need to be seen in a broader perspective. Making East Central Europe secure is part of the larger NATO reform agenda. Done intelligently, enlargement can enhance NATO's overall ability to carry out Article IV missions elsewhere—an important consideration that is often overlooked by critics and advocates alike.

The Baltic issue is also likely to play an indirect role in the debate. While the Baltic countries are not likely to be in the first tranche of enlargement, any attempt to exclude them formally as possible candidates down the road would create internal tensions within the Alliance, particularly with Denmark, which strongly favors Baltic membership. Some Danish officials have even said privately that Denmark would veto enlargement if the Baltic states were formally excluded. Sweden and Finland also might openly oppose NATO enlargement if the Baltic countries were formally excluded from consideration.

The final and most important factor will be the role played by the United States. U.S. leadership will be critical. Both the East Central Europeans and America's European allies will be looking to the United States to provide the political leadership and diplomatic energy to carry out the process of enlargement. If the United States fails to do so, much of the momentum behind enlargement could erode and enlargement could stall. This, in turn, could not only undermine the prospects for the emergence of stability in East Central Europe over the long run, but have a negative impact on U.S. interests in Europe more broadly, provoking a crisis within the

Alliance and opening the way for a serious erosion of transatlantic relations.

At the same time, the processes of NATO and EU enlargement need to be closely harmonized. Conceptually, the two are closely linked. Both are part of the broader process of European enlargement and promoting stability eastward. However, this does not necessarily mean that both processes must occur exactly simultaneously. In some instances, it may be preferable for NATO to enlarge first, since it may take a while for the EU to sort out its internal problems. Poland's security, for instance, should not be tied to whether it can meet the requirements of the Common Agricultural Policy. In other cases, it may be preferable for the EU enlargement to occur first. The main point is that in the end these two processes should be harmonized as much as possible.

BIBLIOGRAPHY

Ananicz, Andrzej, Przemyslaw Grudzinski, Andrzej Olechowski, Janusz Onyszkiewicz, Krzysztof Skubiszewski, and Henryk Szlajfer, *Poland-NATO*, Warsaw: Euro-Atlantic Association and Stefan Batory Foundation, September 1995.

Andorka, Rudolf, "Hungary: Disenchantment after Transition," *The World Today*, December 1994, pp. 233–237.

Barber, Lionel, "Brussels Keeps Shut the Gates to the East," *Financial Times*, November 16, 1995.

Boland, Vincent, "Slovakia Faces Uneasy Future as Reform Slows," *Financial Times*, September 8, 1995.

"Budapests schwieriges Verhältnis zu Bratislava," *Neue Zürcher Zeitung*, November 1, 1995.

"Central Europe: The Winners," *Foreign Report*, October 26, 1995.

"Die Länder Mitteleuropas wollen Aufbau Bosniens und Kroatiens unterstützen," *Frankfurter Allgemeine Zeitung*, October 9, 1995.

Dove, Kevin, and Anthony Robinson, "EBRD Praises "Fast Track" Countries," *Financial Times*, November 2, 1995.

"The EU Goes Slow on Enlargement," *The Economist,* October 28, 1995, pp. 57–58.

"Europäische Union mahnt Slovakei," *Frankfurter Allgemeine Zeitung,* October 27, 1995.

Gebicki, Wojciech, and Anna Marta Gebicka, "Central Europe: A Shift to the Left?" *Survival,* Autumn 1995, pp. 126–138.

Gorka, Sebestyen, "Hungarian Military Reform and Peacekeeping Efforts," *NATO Review,* November 1995, pp. 26–29.

Havel, Vaclav, "New Democracies for Old Europe," *New York Times,* October 17, 1993.

"Hungary Revived by Tough Medicine," *Financial Times,* October 31, 1995.

Interview with Polish Defense Minister Zbigniew Okonski, *Defense News,* November 13–19, 1995, p. 70.

Larrabee, F. Stephen, *East European Military Security after the Cold War,* Santa Monica, Calif.: RAND, MR-254-USDP, 1993.

Maravall, José Maria, and Julian Santamaria, "Political Change in Spain and the Prospects for Democracy," in Guillermo O'Donnell, Phillippe C. Schmitter, and Lawrence Whitehead, eds., *Transitions from Authoritarian Rule: Southern Europe,* Baltimore: The Johns Hopkins University Press, 1986, pp. 93–94.

McNally, Brendan, "Poles Mull Aircraft Swap with Czechs," *Defense News,* August 8–14, 1995.

National Security Bureau, *Security and Defense Strategy of the Republic of Poland,* Warsaw, July 1992.

Szayna, Thomas S., and F. Stephen Larrabee, *East European Military Reform After the Cold War: Implications for the United States,* Santa Monica, Calif.: RAND, MR-523-OSD, 1994.

Tigner, Brooks, "East Europeans Build Regional Defense Ties," *Defense News,* October 2–8, 1995.

"Walesa fordert Verteidigungsminister zum Rücktritt auf," *Frankfurter Allgemeine Zeitung,* October 11, 1994.

Chapter Six

JAPAN

Norm Levin

INTRODUCTION AND BACKGROUND

The end of the Cold War has not only transformed the international environment but stimulated a major process of transition within Japan itself. This transition affects almost all aspects of Japan—political, economic, and social. Neither the slowness in implementing political reform nor continuing difficulties in instituting economic change should obscure the scope and depth of this underlying transition.

Although Japan's security policies are not the central driving issue in the transition, they are affected as well. Like the United States and its NATO allies, Japan has lost the principal security threat on which it predicated its defense policies. A range of other forces is pushing Japanese leaders to define a more active international role for Japan and to articulate a new basis for Japanese defense efforts. Unlike Germany, however, the Japanese have not fully come to terms with their actions preceding and during World War II, and the legitimacy of the military itself has only partially been established. Japan must not only work out new security policies in a radically different environment but must do so in a context of severe political constraints, as well as mounting economic, technical, and demographic pressures.

As the advent of a Socialist-led government suggests, the process of transition has only just begun. Barring a major international crisis, it will take time to work out. In this sense, what has already happened in Japan is less important than what is yet to come. This paper seeks to provide a conceptual framework for viewing the *process* of change as it affects Japanese defense policies and identifying key trends to monitor. The chapter addresses three basic questions:

*   What is important about the past?

*   What has been the more recent effect of the end of the Cold War?

*   What are the issues to watch in the future?

The chapter concludes that significant changes are now possible in Japanese security policies, although radical departures are not likely in the short term and Japan's past will continue to weigh heavily on its longer-term evolution. Among many influences, the role of the United States and the state of U.S.-Japan relations will remain the key determinants of Japan's future direction.[1]

LIVING WITH THE PAST

Change is omnipresent in Japan today. Consider the following:

*   The Liberal-Democratic Party (LDP), the conservative party that dominated Japanese politics for almost four decades, not only agreed to support the chairman of the Socialist Party (SDP), its historic ideological rival and political nemesis, as Japan's next Prime Minister but joined the SDP in a coalition government. The Socialists, in turn, proceeded to publicly renounce virtually every principle underlying SDP policy—the core beliefs that had defined what it means to be a "Socialist" in Japan for over 40 years.

*   Fifteen different parties and political groupings now compete openly for power in Parliament, in place of the de facto "one-and-a-half party" system of the Cold War period.

---

[1]This chapter is explicitly intended to be broad, conceptual, and synthetic. It draws heavily on my previous research (see, for example, Levin et al., 1993, and Levin, 1993a–c). For other recent interpretations, see Fukuyama and Oh (1993), Keddell (1993), Katzenstein and Okawara (1993), Chinworth (1992), and The Edwin O. Reischauer Center for East Asian Studies (1994). This chapter reflects information up until August 1994.

- The average tenure of Japanese governments over the past year has been three months, buffeting Japan's vaunted political stability and increasingly making the country the butt of jokes about becoming the world's first postindustrial banana republic.

Clearly, something important is happening in Japan. At the same time, however, the past continues to plague the process of change in Japanese political, economic, and security policies. In the area of defense, at least three factors from the past continue to influence Japan's debate and help shape prospects for the future.

The first is widespread distrust of the Japanese military and anxiety about potential Japanese "remilitarization." This anxiety is rooted, of course, in Japan's behavior during the 1930s and 1940s, but it is kept alive by Japan's continuing difficulty in coming to terms with its prewar and wartime experience. The aspect of this difficulty most widely discussed is the external one: the minimalist and seemingly grudging quality of Japan's official apologies for its behavior before and during World War II.[2] But there is also an internal dimension: the tendency to dissipate collective responsibility for Japan's behavior by singling out the military as the source of Japanese internal repression and overseas aggression. Both aspects impede the development of a healthy defense establishment, while sustaining foreign suspicions of Japan as an international actor.

All three pillars of Japan's postwar security policies—Article 9 of the Japanese Constitution, the U.S.-Japan Security Treaty, and Japan's Basic Policy of National Defense—were shaped by this distrust and anxiety. Japan's Constitution, which formally renounces war as a sovereign right of the nation and bans both the possession of "war potential" and the use of force as a means for settling international disputes, has been interpreted by successive governments to preclude everything from "offensive" weapons to Japanese participation in "collective" defense. The 1952 U.S.-Japan Security Treaty (revised

---

[2]Former Prime Minister Hosokawa's candid acknowledgment earlier this year of Japanese culpability for its wartime actions was a significant step forward, but it still is far short of Germany's public contrition. Its effect, moreover, continues to be dampened by periodic inflammatory comments by other senior Japanese officials. Meanwhile, at the popular level, computer games like "Commander's Decision"— which rearrange history and omit any mention of Japan's wartime atrocities— continue to proliferate.

in 1960) limits Japan's defense responsibilities to the defense of the Japanese homeland, while codifying Japan's heavy reliance on the United States. And the 1957 Basic Policy of National Defense ensures that Japan's defense buildup would be modest ("in accord with national capability and the domestic situation"), gradual, and "within the limits necessary for self-defense," while making clear that the U.S.-Japan Security Treaty (i.e., the United States) would serve as the "basis" for Japan's defense. All three continue to structure and circumscribe Japanese defense policies today.

The second factor relates to the origin and evolution of the Self-Defense Forces (SDF). Blamed for Japan's lurch toward militarism in the 1930s and disastrous defeat in World War II, officially relegated to a secondary status in state councils (the SDF is lodged in a Defense *agency*, not ministry, and the head of the forces neither commands troops nor is permitted to testify in the Diet), and constrained against either defining military requirements in terms of the capabilities of any particular foreign state or positing military roles that exceed the strict territorial defense of the Japanese homeland, the SDF grew up as a foreign step-child in a family seeking international assimilation. A short summary of some of the major ramifications of these unique constraints includes the following: a defense buildup not rooted in any coherent strategic doctrine; a uniquely defensive orientation in operational concepts, training, and planning; a lack of balance in mission areas, especially concerning force projection and offensive operations; and perennial difficulties in joint planning and interservice relations.

Many of these characteristics are evident in the 1976 National Defense Program Outline (NDPO), the principal modification of Japan's fundamental defense policies developed in the 1950s. The NDPO identified an ability to counter "limited and small-scale aggression" as the target of Japan's defense buildup, without defining what such an aggression might look like or how it might occur. It articulated a vague military strategy of denial or prevention—hindering an aggressor from accomplishing his objective and sustaining resistance until outside assistance could arrive—to repel such aggression. And it defined as the "basic defense capability" Japan would maintain to achieve this strategy a military force structure roughly comparable to what existed at the time.

Implicit in the NDPO is a general Japanese difficulty in envisioning military conflict involving Japan in anything other than global terms. Given Japan's objective dependence on the United States and growing uncertainties (in the wake of Vietnam) about timely U.S. assistance, any such conflict would require significantly expanded Japanese defense roles and capabilities as part of a more coordinated defense with the United States. The principal roles of the SDF gradually came to reflect this implicit recognition, with new divisions of labor between Japan and the United States: the Air Self-Defense Force (ASDF) would provide air defense as part of U.S. strike packages, the Maritime Self-Defense Force (MSDF) would protect U.S. carrier battle groups and defend southern sea lines of communication, and the Ground Self-Defense Force (GSDF) would assume full responsibility for protecting Japanese territory against potential landings. Despite the emphasis on "limited and small-scale aggression" as the target of Japan's defense buildup, the NDPO actually allowed for significant quantitative and qualitative improvements in SDF capabilities to fulfill these strategic objectives. As a result, 18 years after the NDPO's adoption, the SDF has become a relatively modern, medium-sized force with some of the world's most sophisticated military equipment. But it has not been able to meet the numerical targets identified in the NDPO, particularly in the case of the Ground Self-Defense Force, and continues to suffer from major force structure and operational shortcomings.

The third factor is heavy dependence on the United States. As noted above, this dependence is built into Japan's fundamental defense policies. It is reinforced by Japan's inherent geostrategic vulnerabilities—Japan is an island state surrounded by either large continental powers or vast expanses of ocean; it is dependent for its survival on imports of oil and other natural resources from distant places; its large population is heavily concentrated in a small geographic area; and it has no natural regional allies—and a host of other self-imposed constraints on developing the ability to unilaterally deter and defend against hostile powers (such as the ban on nuclear weapons). This objective dependence has made the United States the core of Japanese foreign and domestic policies for over 40 years.

This heavy dependence has had mixed consequences for the SDF. On the one hand, close interactions with the United States have helped to improve operational capabilities significantly in recent years, particularly regarding the MSDF and ASDF. The SDF has also been freed to focus on Japan's territorial defense requirements, while gaining access to important defense technologies that might not otherwise have been available. On the other hand, close ties with the United States have affected the SDF's ability to operate autonomously, particularly in the case of the MSDF, which has historically tied its force structure and mission planning most closely to the United States. Precisely how important an ability to operate more "autonomously" is remains a critical question for future resolution.

These factors from the past have contributed to giving Japan's defense its striking characteristics: a political establishment that places top priority on maintaining close relations with the United States; a medium-sized military force (which, while equipped with some of the most advanced equipment, has serious operational and other shortcomings, is consistently unable to meet its force planning targets, and remains oriented to territorial defense as a junior partner of the United States); and a public that remains suspicious of the military and wary of letting the military "genie" out of the bottle.

LIVING WITHOUT THE SOVIETS

For several years, the effect of the Soviet Union's collapse and the end of the Cold War was muffled in Japan. The historic Japanese suspicion of Russia and the unresolved dispute over Japan's "northern territories," coupled with the continued presence of large numbers of well-equipped Russian forces in areas around Japan, reinforced Japanese doubts over the permanence of change in Russia and bolstered the defense establishment's reluctance to modify Japan's security policies significantly. The continued Cold War stalemate in Korea and the perseverance of communism in Asia despite developments in the former Soviet Union and Eastern Europe further retarded Japanese grappling with the reality of a new era.

Despite the delay in Japanese policy adjustments, the end of the Cold War has had profound effects, which have become increasingly visible over the past year. Perhaps the most obvious is an undermining of the central logic on which Japan's defense plans were predicated. The NDPO, as noted above, was predicated on the notion of Japanese defense against "limited and small-scale aggression."

Although Japanese officials publicly explained that such conflict was the only kind Japan would be likely to face because the Security Treaty would deter any larger forms of aggression, this notion in fact reflected the belief that any major attack on Japan would almost necessarily be part of a larger global conflict. Japan's responsibility in such a conflict would be to "hold out" for a period of time against attacks on Japan and prevent any *faits accomplis* until the arrival of U.S. military assistance. Although political sensitivities prevented the government from formally designating the Soviet Union as Japan's enemy, the object of Japan's defense buildup was clear. And because of the dramatic Soviet military buildup of the 1970s and early 1980s, Japan's midterm defense plans formulated on the basis of the NDPO not only provided for significant annual increases in defense spending but focused in particular on heavy purchases of advanced, front-line equipment. The demise of the Soviet Union as a credible threat and the decreased likelihood of global conflict undermined this central logic and shifted Japanese threat perceptions to a more variegated set of concerns focused on dangers *within* Asia. It also stimulated demands for revision of the NDPO itself, particularly its attached table of organization, which was predicated on a growing threat from the former Soviet Union.

The loss of the Soviet threat has also affected the SDF, particularly the GSDF. The MSDF has always been more concerned with protecting Japan's long sea lines of communication than in resisting some presumed Soviet invasion of the homeland and has focused its attention not on the north but on helping protect U.S. naval forces and securing Japan's sea lanes to the south and southwest. Similarly, the ASDF's principal role of protecting Japanese airspace has not fundamentally been called into question, although the rationale for such a large number of advanced (and expensive) aircraft absent the Soviet air threat is increasingly challenged. But without the Soviet Union—the only country arguably capable of landing troops in Japan—both the size and fundamental *raison d'être* of the GSDF has been called into question. Unable to come close to meeting its authorized troop strength in the best of times, the ground forces have borne the brunt of demands for force structure and operational planning modifications.

A third and related effect has been to intensify downward pressures on defense spending. After breaking the 1 percent of GNP barrier in

1987, Japanese defense expenditures declined as a share of GNP by nearly 10 percent (from 1.013 percent to 0.937 percent) between 1988 and 1993. The 1.2 percent rate of increase in defense spending in 1993 was the lowest rate of growth since 1960 and a striking reduction from average annual increases in excess of 5 percent throughout the 1980s. The rate of increase in defense spending declined further in fiscal year 1994 to 0.9 percent, equivalent to a real cut of at least 1 percent. These trends reflect a government decision in 1992 to revise the midterm defense plan one year ahead of schedule and to lower significantly both the plan's total programmed defense expenditures (by about 580 billion yen) and average rate of growth (from 3 percent to 2.1 percent in real terms annually) (*Boei Nenkan*, 1993, pp. 176–177). As Table 6.1 suggests, these downward pressures are especially strong on spending for front-line equipment. Contract expenditures for such procurements were reduced in the revised midterm plan by about 5 trillion yen, with major cutbacks in the planned procurement of tanks, destroyers, and fighter aircraft.

Another effect of the end of the Cold War has been to heighten public awareness of the need for greater Japanese contributions to international security. The roots of this awareness actually go back a long way. The oil crisis in 1973 and U.S. experience in Vietnam in the mid-1970s awakened the Japanese to the effect of developments abroad on Japan's own security and marked the beginning of a new Japanese interest in closer military cooperation with the United States. The Soviet invasion of Afghanistan in late 1979 and subsequent intensification of the Cold War in the early 1980s precipitated

Table 6.1

Comparison of Main Equipment Procurement Under Midterm Defense Plan

Type	Defense Plan (1986–1990)	Initial Midterm Defense Plan (1991–1995)	Revised Midterm Defense Plan (1991–1995)	Changes from Initial to Revised Plan
Tanks	246	132	108	−24
Destroyers	9	10	8	−2
Fighter aircraft	63	42	29	−13

SOURCE: Japan Defense Agency (1993).

a new focus on Japan's role "as a member of the West" and an incipient move—through increased Japanese support for Western political objectives, significantly expanded economic assistance to countries of strategic importance to the West, etc.—toward greater international activism. The combination of mounting Japanese economic strength and growing U.S. economic difficulties more broadly convinced many Japanese by the end of the decade that they could no longer behave like the tall schoolboy who slouches in the back of the classroom hoping no one will notice his presence.

But the end of the Cold War heightened this awareness in a number of ways. Probably most important, it diminished the sense of Japan's importance in U.S. global strategy and, in the context of America's ongoing domestic difficulties, increased U.S. pressure for greater allied burden-sharing. By making it clear that a continued heavy American security role would rest on increased responsibility-sharing by U.S. allies—as demonstrated graphically in Desert Storm—the end of the Cold War provided Japan incentives to explore ways to expand its international contributions seriously. At the same time, it posed a clear threat of international isolation, perhaps Japan's single greatest historical fear, if greater Japanese contributions were not forthcoming. Again, the impact of Desert Storm on Japanese attitudes was enormous. Finally, the end of the Cold War provided opportunities and potential rewards: With the Soviet Union no longer around to veto UN actions, achieving Japan's long-sought objective of a permanent seat on the Security Council now became a distinct possibility—but only if Japan could demonstrate that it is prepared to support UN activities fully. Such considerations precipitated the major new development in Japanese security policies since adoption of the NDPO: passage of legislation in June 1992 enabling the SDF to be sent overseas and Japan's subsequent first participation in international peacekeeping operations (in Cambodia and Mozambique).

Three other, more indirect effects of the Cold War's end should also be noted. First, it weakened the glue that has helped sustain U.S.-Japan relations through two decades of mounting economic tensions. On the American side, as reflected in the Clinton administration's "results oriented" trade policy, the absence of an overarching threat has bolstered the tendency to give greater priority to narrow U.S. economic interests. On the Japanese side, the new partnership between Russia and the United States reinforced anxieties about Japan's place in U.S. global strategy, particularly given continuing Japanese concerns about Russian military power deployed in the Far East and potential Russian pressures. On both sides, incipient but perceptible signs of a growing indifference to each other suggest the possibility of a gradual mutual disengagement—not a "pulling" but a "drifting" apart fed by the domestic political situations in each country. While Washington and Tokyo have officially stressed the continuing strength of U.S.-Japan political and security ties despite the heightened strains in the bilateral economic relationship, the Japanese have grown increasingly nervous about their ties with the United States and open to considering new policy departures. One result has been a distinct increase in the relative importance placed on Asia in Japanese diplomacy.

Second, the end of the Cold War reinforced the historic Japanese sense of weakness and vulnerability. This may seem somewhat counterintuitive: One might assume that the collapse of the largest potential threat to Japan would have precisely the opposite effect, and certainly the strong downward pressures on defense suggest a more relaxed national psyche. But while the Soviet Union is gone, Russia remains, as does a major territorial dispute that prevents normalization of Russian-Japanese relations. Along with continued concern about a latent Russian threat, moreover, have come new anxieties about Japan's other neighbors, especially North Korea and China, whose futures are increasingly uncertain and with whom Japan has experienced difficult historical relations. At the same, the end of the Cold War has called into question the long-term staying power of the United States and the utility of the American security guarantee against non-Soviet threats to Japanese security. The sustained Japanese recession and protracted political turmoil reinforce the sense of limitations on Japan's ability to cope with the exigencies of the new international situation. In the short term, this bolsters Japanese incentives to shore up the U.S.-Japan security relationship to limit Japan's exposure.

Finally, the end of the Cold War facilitated the initiation of major political realignment in Japan. Clearly, the collapse of the LDP and the rise of a new coalition government cannot be attributed directly to the global situation. Systemic corruption, power struggles within the LDP, and a general Japanese desire for change were the central causes. But the ending of the Cold War removed the ideological component of domestic Japanese political conflict and weakened the LDP's claim to a unique status as the guardian of Japan's national security. It also exacerbated long-standing fissures within the major political parties themselves and provided a basis for new groupings along personal and policy lines. In the short term, the process of political realignment will almost surely involve a period of fluidity and weak leadership, minimizing the prospect for any radical changes in Japanese security policies. Over the longer term, however, the new alignments set off in part by the advent of a new international era could pave the way to significant departures.

DEBATING DEFENSE AFTER THE COLD WAR

Not surprisingly, these developments have influenced the nature of the current defense debate. At the broadest level, the combination of Japan's new security environment, political and other constraints, and continuing geostrategic vulnerability have lowered the salience of two polar schools of thought that dominated Japan's defense debate in the first two decades of the postwar period: unarmed neutrality (as long promulgated by Japan's leftist Socialist Party) and independence (in the Gaullist sense). The former has been undermined by the diminished credibility of its principal proponents, as well as by its inherent lack of realism. The latter has been weakened—in an age of North Korean long-range missiles and political turmoil in Tokyo—by growing appreciation of Japan's limitations and objective dependence on the United States. While they remain the principal schools of thought advocating *no* reliance on the United States, the principal debate today is over two alternative visions: the "Japan as a normal country" school of thought, most commonly associated with long-time LDP strongman and current Shinseito (Japan Renewal Party) leader Ichiro Ozawa, and the "Japan as a global civilian power" orientation most directly associated with Masayoshi Takemura, leader of the Shinto Sakigake (New Party Harbinger). Both schools of thought go beyond security issues per se to address basic political and economic issues, but they have important defense components as well.

Proponents of transforming Japan into a "normal country" draw from a long-standing Japanese school of thought that seeks a more assertive international role for Japan. In some respects, the "normal country" philosophy may be considered essentially an updating and further extension of the "autonomous defense" school of thought, which, while present even in the 1950s, became particularly prominent in the late 1960s and thereafter because of its association with Defense Agency Director-General and later Prime Minister Yasuhiro Nakasone. Leaders of both schools of thought share several key beliefs in common: They believe that Japan as a matter of principle should assume primary responsibility for its own defense, as well as its "rightful" place as a regional and global leader; they consider the external environment as inherently dangerous and regard notions like "limited and small-scale aggression" as nonsensical goals for Japan's defense buildup in the real world; they see the constitutional constraints on Japanese defense efforts as both unnatural and an insurmountable obstacle to the development of a meaningful self-defense capability; and, unlike Japan's Gaullists, they consider close U.S.-Japan security relations to be absolutely essential to Japan and want to expand Japanese roles and responsibilities in the alliance to ensure its perpetuation. Based on these views, advocates of this school of thought want to revise, or at least reinterpret, the Constitution to further legitimize the SDF, promote military cooperation with the United States, and allow greater Japanese participation in UN and international collective security operations.

Those whose vision of Japan is as a "global civilian power" also draw on a long-standing intellectual and philosophical tradition, one that regards the principles of the "peace constitution" as embodying contemporary Japanese uniqueness and a policy based on these principles as the way to fulfill Japan's nationalist aspirations. While they agree that Japan has an important role to play in the international community, they see this role as being essentially nonmilitary. They also see a far more benign security environment than do their principal antagonists and, in any event, tend to downplay the utility of military force for solving international problems. Like their "normal country" counterparts, they agree that ties with the United States are critical to Japanese interests and want to maintain close security relations. But they seek greater equality in Japanese-American interactions and greater balance in Japan's foreign relations more broadly. It may be no coincidence in this regard, as suggested in Figure 6.1, that the most prominent representative of this

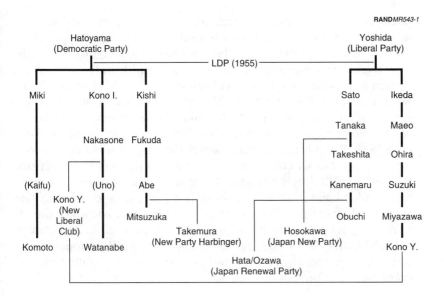

Figure 6.1—Genealogy of Main Japanese Conservative Factions

school of thought, Masayoshi Takemura, emerged from an LDP faction that traces its roots back to former Prime Minister Kishi, who pushed to revise the Security Treaty with the United States to make it more "equal." Kishi's successor as faction head, former Prime Minister Fukuda, promulgated the "Fukuda Doctrine" during his tenure, which called for greater emphasis on Asia and an "omnidirectional" foreign policy.

Today, advocates of the "global civilian power" school of thought see opportunities in the post–Cold War era to decrease military spending and allocate greater resources to public welfare and international assistance. They thus support cutbacks in the size and strength of the SDF while maintaining Japan's fundamental defense policies. Proponents of this school actively support the UN, but they are cautious about (and many are opposed to) Japan's seeking a permanent seat on the Security Council because they fear this would require Japanese military participation in UN-led operations.

As a general statement, the "Japan as a normal country" school finds its main proponents in the Shinseito (Japan Renewal Party), Nihon Shinto (Japan New Party), and the center-right and right wings of the LDP (Nakasone-Watanabe faction in particular). Advocates of the "Japan as a global civilian power" school tend to be clustered in the Shinto Sakigake (New Party Harbinger), the mainstream (center and center-left) of the LDP, and the Socialist Party (although the left side of the party adheres to an even more restrictive and nonaligned orientation). But it is important to stress that supporters of one versus another of these schools may be found across the political spectrum, and party affiliation is not a reliable guide to views on security matters. Indeed, clear identification of one party or another with particular policy views is likely to require further breakdown of the old party system and a new, stable political realignment. As is evident in the SDP-LDP union, current political alignments reflect opportunism and search for political power, not coalitions formed on the basis of common views on security issues.

Still, positions on the underlying debate between these two schools of thought, coupled with a broader desire to maintain many elements of the status quo, provide at least some measure of commonalty among the members of the current government coalition. Along with a deeply shared personal antipathy toward the leader of the "normal country" school, Ichiro Ozawa, this commonalty could provide the coalition a longer tenure than has generally been expected.

Whatever the ultimate prospects for political realignment, two fundamental issues lie beneath the debate between these broad schools of thought: what kind of military Japan should maintain in the new security environment and how Tokyo should structure its relations with the United States. These questions have been the subtheme of much of recent security debate.

The first question subsumes a broad range of issues, but debate has centered on two in particular

Revising the NDPO

This has probably been the greatest subject of debate over the past year and contains a certain irony. When the issue first arose back in the mid-1980s, the pressure for revision came primarily from those alarmed by the rapid Soviet military buildup and the growing gap between Soviet strength and Japan's ability to respond with the modest goals laid out in the NDPO. With the end of the Cold War, the collapse of the USSR, and the development of a new Russian-American relationship, the tables turned, and critics of expanded Japanese defense efforts became the principal advocates of revision.

The main focus of debate has been over the program's attached table, summarized in Table 6.2, identifying the maximum authorized strength for each of the three services. Critics charge that this table

Table 6.2

NDPO Attached Table

Force	Units and Equipment	Quantity
GSDF	Total personnel	180,000
Basic units	Regionally deployed units	12 divisions
		2 combined brigades
	Mobile operation units	1 armored division
		1 artillery brigade
		1 airborne brigade
		1 training brigade
		1 helicopter brigade
	Ground-to-air missile units	8 artillery groups
MSDF		
Basic units	Mobile surface ship units	4 escort flotillas
	District surface ship units	10 divisions
	Submarine units	6 divisions
	Minesweeping units	2 flotillas
	ASW aircraft units	16 squadrons
Main equipment	ASW surface ships	60 ships (approx.)
	Submarines	16 submarines
	Combat aircraft	220 aircraft (approx.)
ASDF		
Basic units	Air control and warning units	28 groups
	Interceptor units	10 squadrons
	Support fighter units	3 squadrons
	Air reconnaissance units	1 squadron
	Air transport units	3 squadrons
	Early warning units	1 squadron
	Ground-to-air missile units	6 groups
Main equipment	Combat aircraft	430 aircraft (approx.)

SOURCE: Japan Defense Agency (1993).

provides a force structure that is both excessively large for the needs of the post–Cold War era and increasingly hollow in terms of real fighting capabilities (given the emphasis on procuring front-line equipment and the relative lack of attention to spare parts, ammunition, and other support capabilities). With the large gap between its authorized and actual manpower levels and irreversible (at least in the short to middle term) demographic trends, the GSDF has been under particularly strong pressure to reduce and restructure its forces. Even well-known defense specialists suggest the ground forces could make do with perhaps a force of 130,000 (as opposed to its authorized level of 180,000 and current strength of around 150,000), with some calling for a force as small as 100,000 divided into perhaps five divisions (as opposed to the present 12 divisions and two combined brigades). Budgetary constraints are putting pressure on the ASDF and MSDF as well, however, as the sharp debate over procurement plans for airborne warning and control aircraft, P-3Cs, and F-15s makes evident.

At the same time, critics have assailed the basic philosophy underlying the NDPO, particularly its notion of a single major security threat to Japan and the concept of a "basic defense capability" to supplement the U.S.-Japan Security Treaty in resisting such aggression. This philosophy dictates an emphasis on the buildup of a range of sophisticated equipment and weapon systems—F-15s, airborne warning and control aircraft, P-3Cs, Aegis-equipped destroyers, etc.—that critics claim are not appropriate priorities for Japan's relatively limited security needs. Together with the 1978 Guidelines on U.S.-Japan Defense Cooperation, which provided the basis for significantly expanded cooperation for dealing with threats in the Far East, the basic philosophy of the NDPO effectively pushes out the range of Japan's defense responsibilities and expands its cooperation militarily with the United States.[3] The issues of whether the SDF of the future will be optimized for joint operations with the U.S.—including questions of Japanese efforts on issues ranging from cross-servicing arrangements to multinational peacekeeping activities—or whether

---

[3]These criticisms join long-standing complaints within the Japanese military about the NDPO's basic philosophy. Senior officers have long seen this philosophy as militarily illogical and have harshly criticized both the underlying concept and buildup approach.

future buildup plans will emphasize more narrow, autonomous operations are tied up in this underlying debate. At least a tentative answer may be available this fall after the government and private advisory committees appointed by former Prime Minister Hosokawa issue their respective recommendations.

Participating in Peacekeeping and Other Overseas Operations

Passage of legislation to allow Japanese participation in UN-led international peacekeeping operations and the dispatch of SDF members to Cambodia and Mozambique broke a long-standing taboo against Japanese military activities. It did not end the debate, however, over Japan's involvement in peacekeeping, rescue, and other types of overseas operations. Not only is the legislation riddled with restrictions (e.g., the total number of Japanese personnel may not exceed 2,000; any SDF participation must be confined only to rear support duties; SDF personnel may not be equipped with anything other than side arms for personal self-defense; etc.), but every instance of proposed SDF participation must receive formal Diet approval. Debate continues, moreover, over whether Japanese participation in peacekeeping operations, even if mandated by the UN, violates Japan's constitution and whether the UN charter or the Japanese constitution takes precedence. Precisely for this reason, those of the "Japan as a normal country" school are pushing for changes in the provisions or interpretation of the constitution that make clear Japan's ability to support such UN-led activities. Proponents of the "global civilian power" school are strongly resistant.

In an effort to tone down debate, former Prime Minister Hosokawa straddled the issue in his 1993 address to the UN General Assembly by stating that Japan would honor its UN responsibilities "to the extent possible," enigmatically adding the further qualifier "in a reformed United Nations." Meanwhile, debate has swirled over whether the SDF can be sent overseas even to rescue Japanese nationals caught in dangerous situations without first revising laws governing the SDF. The extent to which Japan will acquire the military capabilities to be able actually to participate extensively in such activities, as suggested above, is also uncertain.

The second question, how Japan should structure its security relations with the United States, has been much less a focus of attention, reflecting in part the broad consensus that has developed over the past 10 to 15 years on the importance of the U.S.-Japan security connection. Two issues, however, are the subject of particular debate.

Responding to North Korea

This issue has been a divisive one in Japan throughout the postwar period, as reflected in sharp debates over Japanese support of U.S. military roles during and after the Korean War, normalization of relations with South Korea (in response to strong U.S. pressure) in 1965, and the Nixon-Sato joint communiqué in 1969 publicly affirming the importance of Korea to Japan's security. But North Korea's nuclear and missile development programs and the incipient U.S.-led push for international sanctions against Pyongyang have raised the salience of this issue once again.

The current debate revolves around Japan's support for, and participation in, three Korea-related activities: international sanctions, a theater missile defense system, and potential U.S. military operations. As a general statement, the advocates of Japan as a "normal country" tend to support active Japanese participation in all of these activities, should they become necessary. Proponents of the "global civilian power" school range from reluctant to strongly opposed. But there is a range of views within each school on all of these activities. What is important to stress is the power and volatility of the overall issue. Perhaps alone among the major issues involved in the process of political realignment, Korea has the potential to become a defining issue in Japan's ultimate domestic evolution. Should a crisis flare up on the Peninsula requiring U.S. military actions, for example, the present government would almost surely collapse as the two main parties in the coalition (the LDP and SDP) splintered and a new amalgamation of centrist-conservative forces formed to protect Japan's *national* interests. Conversely, a U.S. failure to respond adequately to a crisis would fundamentally alter Japan's central strategic calculations and affect both the nature and role of future Japanese governments.

Japanese nuclearization is not a prominent aspect of the current debate over Korea. Public opinion polls consistently demonstrate that the overwhelming majority of Japanese remain strongly opposed to the possession of nuclear weapons, and few Japanese political leaders publicly advocate their acquisition. Indeed, antinuclear sentiment is so strong in Japan that even the known existence of nuclear weapons on the Korean Peninsula would not guarantee that Japan itself would "go nuclear." The Japanese government's decision in fall 1994 to formally endorse an indefinite extension of the nuclear Non-Proliferation Treaty when it expires in 1995—despite mounting concern with North Korea—reflects its awareness of this strong public sentiment.

Having said this, Japanese governments have made clear since the 1950s that Japan's nonnuclear stance is merely a *policy* of the particular government that endorses it and cannot bind future Japanese governments. Japan has worked assiduously, moreover, to make sure that it has the technical and economic potential to manufacture nuclear weapons should this be deemed necessary. With its ambitious civilian nuclear energy program, large and growing supply of plutonium, and technical capabilities in precision machining and other nuclear weapon-related industries, there is little doubt that Japan could produce nuclear arms relatively quickly (although it would, of course, first have to withdraw from the treaty) if it chose to do so (Williams, 1994). Korea is unique among current and foreseeable issues in having the potential ability to precipitate a sea-change in Japanese thinking. Already, it has broken the long-standing taboo against *discussion* of possible Japanese nuclearization.[4]

Sharing Technology

This issue dates back to the early 1980s when the United States first began to seek a more active two-way flow of technology as part of a more equitable security relationship. The Japanese decision in 1983 to make an exception to the government's ban on arms exports and allow the transfer of military and dual-use technology represented a milestone for Japan, and was quickly followed up by Japanese agreement to the establishment of a Joint Military Technology

---

[4]For the seriousness with which the defense establishment views the Korea issue in general and North Korean nuclear and missile development activities in particular, see Japan Defense Agency (1994).

Commission to identify appropriate technologies for possible transfer. The record on successful transfers over the past decade, however, has been spotty at best, and new U.S. pressures for access to key Japanese military and especially dual-use technologies (e.g., the Perry Initiative) have once again increased the salience of this issue.

Technology sharing raises a complex set of issues, most of which are between the respective private sectors and between the Japanese government and private sector. Public debate stems from two sources: suspicions in certain quarters in Japan that the U.S. request is designed not to improve military capabilities but to raid competitive Japanese civilian technologies and political sensitivities about violating the government's restrictions on arms exports and the constitutional injunction (as a matter of interpretation, not law) against participation in "collective" security activities. The Theater Missile Defense Initiative (TMDI) raises the additional issue of potential violations of the spirit of the UN resolution limiting the use of space to peaceful purposes, which the Japanese Diet adopted in 1969. Sensitivities on these matters are particularly strong within the Socialist Party, although they are present in the LDP and other parties as well. As long as the government depends on political support from these sectors, making rapid progress on TMDI and similar technology-sharing issues will be difficult.

CONCLUSION

This review of the current defense debate suggests several key trends and issues to watch carefully. One is Japan's shifting attitudes about Korea. Another is the evolution of thinking about Japan's next-term defense plan and the nature and scope of Japanese defense requirements in the post–Cold War era. A third is the direction of debate over Japanese participation in peacekeeping, rescue, and other overseas operations. Any of these issues could significantly affect future Japanese defense policies and prospects for U.S.-Japan security cooperation. More broadly, trends affecting the further splintering of the major political parties (especially the LDP and SDP) need to be monitored closely, since these will not only help determine future political alignments but—as the debate between the "normal country" and "global civilian power" schools of thought suggests—the future role of Japan as a national actor.

Stepping outside the defense debate itself, a number of broader generalizations are possible. The following should be kept in mind as Japan continues with its domestic transition:

- *The name of the game today is political survival, not national policy.* While ideas are fundamentally at the heart of Japan's current political turmoil, the series of ad hoc alignments and the rapid rise and fall of governments reflect essentially opportunistic maneuverings to cope with a system in transition. Some time will be required before a stable balance—let alone a new party system—can be established. Until then, there will be only a loose, if any, correlation between political coalitions and specific policy positions.

- *The absence of agreement on fundamental defense issues makes rapid policy movement unlikely.* Neither the previous "reform" governments (Hosokawa and Hata) nor the present SDP-LDP coalition established a basis for agreement among the coalition partners on basic security policies. This is true on issues affecting both Japan's indigenous defense buildup and U.S.-Japan security cooperation. Until greater consensus can be achieved, we should not expect either radical departures in Japanese security policies or rapid progress on issues of concern to the United States (e.g., TMDI).

- *The weakening of the political extremes and expansion of the middle both narrow the range of policy options and increase the possibility of significant change over the longer term.* Like international politics, Japanese politics were largely "deideologized" by the collapse of the Soviet Union and the end of the Cold War. It is precisely this deideologization that made the recent SDP-LDP coalition possible. The implications for Japanese defense policies are twofold: In the short term, the sharp polarization that has characterized Japan's defense debate will be attenuated and policies will fluctuate within a range supportable by the broad "middle." Over the longer term, however, significant departures in Japanese policies will be possible as a stable power balance is established and a new policy consensus develops. Before such departures would involve a "renationalization" of Japanese security policies, at least three conditions would need to be met: a rupture in U.S.-Japan relations, the rise of a major external security threat, and the development of a genuine sense of Japanese isolation and a new public consensus behind a more truly "independent" security posture.

- *It is important to distinguish between the constant and the ephemeral.* Whatever happens on the political front, certain trends—such as a declining birth rate, the escalating cost of new weapon systems, and decreased defense procurement and R&D because of the need to both raise the wages and improve the working conditions of servicemen—will continue to put pressures on Japanese defense policies. In the absence of a new regional crisis, a slowdown in Japan's defense buildup, and perhaps even downsizing of Japan's military, is almost inevitable.

- *The single greatest determinant of future Japanese defense policies remains the United States.* Among the constants alluded to above is the centrality of the United States to Japan's fundamental national interests. Almost all major political groupings except the communists now accept this proposition. Barring a collapse of the U.S. security commitment or rupture in U.S.-Japan relations, future changes in Japanese defense policies will take place within the context of Security Treaty arrangements. But we should not treat lightly indications of an underlying disquiet: growing Japanese uncertainties over the wisdom of U.S. policies toward North Korea; unease over Washington's optimistic appraisal of Russia's prospects and, from a Japanese perspective, insensitivity toward Japan's interests; questions about Japan's ability to rely on the United States to remain an effective counterbalance to growing Chinese military capabilities; and continuing doubts about U.S. long-term staying power in the region more broadly. Together, these represent symptoms of a security relationship that is not as healthy as both sides officially profess and constitute potential spurs to more radical Japanese policy departures. Continued close U.S.-Japan security cooperation in this environment will require both continued U.S. engagement and even greater high-level attention to Japan. It also will require articulation of a clear rationale for U.S.-Japan security relations in the post–Cold War era and an ability to fit U.S. economic objectives into a larger vision of U.S. strategic interests.

REFERENCES

Boei Nenkan, 1993, pp. 176–177.

By the Way, January-February, 1994.

Chinworth, Michael, *Inside Japan's Defense,* New York: Brassey's Inc., 1992.

The Edwin O. Reischauer Center for East Asian Studies, *The United States and Japan in 1994: Uncertain Prospects,* Washington, D.C.: Nitze School of Advanced International Studies, 1994.

Fukuyama, Francis, and Kongdan Oh, *The U.S.-Japan Security Relationship After the Cold War,* Santa Monica, Calif.:   RAND, MR-283-OSD, 1993.

Japan Defense Agency, *Defense of Japan, 1993,* Tokyo, 1993.

_____, *Defense of Japan, 1994,* Tokyo, 1994.

Katzenstein, Peter, and Nobuo Okawara, *Japan's National Security,* Ithaca, New York: Cornell University East Asia Program , 1993.

Keddell, Jr., Joseph, *The Politics of Defense in Japan: Managing Internal and External Pressures,* London: M. E. Sharpe, Inc., 1993.

Levin, Norman D., "Japan's Defense Policy: The Internal Debate," in Harry Kendall and Clara Joewono, eds., *Japan, ASEAN, and the United States,* Berkeley: University of California Press, 1993a.

_____, "Prospects for U.S.-Japanese Security Cooperation," in Danny Unger and Paul Blackburn, ed., *Japan's Emerging Global Role,* Boulder, Colo.: Lynne Rienner Publishers, 1993b.

_____, "The Strategic Dimensions of Japanese Foreign Policy," in Gerald Curtis, ed., *Japan's Foreign Policy,* London: M. E. Sharpe, Inc., 1993c.

Levin, Norman D., Mark Lorell, and Arthur Alexander, *The Wary Warriors: Future Directions in Japanese Security Policies,* Santa Monica, Calif.: RAND, MR-101-AF, 1993.

Williams, Michael, "Japan Was Urged to Keep Potential for Nuclear Arms," *The Wall Street Journal,* August 2, 1994, p. 9.

Chapter Seven

KOREA

Abram N. Shulsky

In the past year, Korea, especially the North Korean nuclear issue, has been one of this country's most urgent foreign policy concerns. While this chapter is not the place for a detailed review of the current situation, it will attempt to place it in a somewhat broader and longer-term perspective, and to look at some issues that could emerge once the issues which currently concern us have been surmounted.

UNDERSTANDING THE CURRENT SITUATION ON THE KOREAN PENINSULA

One of the key problems in understanding the current situation is assessing the expected longevity of the North Korean regime—either as it exists now or as it might be transformed if it were to "open" itself to the world economically—and what role its nuclear program plays in its survival strategy.

In the midst of an economic crisis brought on primarily by Russian and Chinese unwillingness to continue to subsidize its economy through barter trade on favorable terms, the North Korean regime appears to be a curious combination of strength and vulnerability. The state's repressive apparatus remains intact, and it appears capable of moving forward on high priority objectives, such as its nuclear program. The transfer of power from the late Kim Il-Sung to his son, Kim Chong-Il, appears to be on track, although questions have arisen about the latter's health, suggesting that he may be a "transitional" leader. It is important to remember, however, that those selected to be "transitional" leaders, such as Leonid Brezhnev, often have a way of hanging around far longer than anyone expected. In any case, the relative smoothness of the succession process, despite Kim Chong-Il's lack of his father's charismatic personality and his reputation as something of a playboy, would seem to demonstrate that the basic apparatus of power remains strong.

On the other hand, the economic crisis is severe, as the regime publicly admitted last year. In addition, reports of strong popular dissatisfaction and even food riots, as well as rumors of unrest in the military, surface from time to time. The shock of seeing almost all communist regimes around the world fail within several years must have been a severe one. The most important remaining communist government, that of China, established diplomatic relations with South Korea and is busily increasing its economic ties as well.

Despite appearances, the North Korean regime may nevertheless be in the midst of a delayed succession crisis following the death of Kim Il-Sung. Even if Kim Chong-Il succeeds to all his father's posts, it may be that forces opposed to him are merely biding their time, perhaps in the belief that his ill health will force him to step down in a few years in any case. Alternatively, they may be unready to make their move now, but may feel that, if the younger Kim is in fact unable to grab all the reins of power, the opportunities for intrigue will be greater in the future. However, it must be acknowledged that our insight into the inner workings of the regime is extremely limited; we are as likely as not to be quite surprised by the way the succession works itself out.

In the wake of the agreement[1] signed between North Korea and the United States, it is possible, however, to get a better sense of how the

---

[1]Under this agreement, the United States is to organize an international consortium that will provide North Korea with two light-water reactors (LWR) and, pending completion of the first unit, with shipments of heavy oil. North Korea agrees to freeze and, on completion of the LWR project, dismantle its graphite-moderated reactors and (undefined) "related facilities." The United States and North Korea are to cooperate in finding a method to store safely and eventually to dispose of the spent fuel rods from the current reactor "in a safe manner that does not involve reprocessing in the DPRK." (U.S. spokesmen have interpreted this phrase as implying the shipping of the fuel out of North Korea.) The Democratic People's Republic of Korea (DPRK) will allow

North Korean leadership intends to surmount its economic crisis. Rather than "cash in" its nuclear program in return for economic benefits or move forward aggressively toward acquiring nuclear weapons (or expanding its arsenal, if it already possesses one or two), North Korea is following a policy of eating its cake and having it, too. In other words, it intends to preserve the option of pursuing a basic nuclear capability, while selling concessions that fall short of closing off that option for the highest price it can get. The current agreement puts off for at least five years any step North Korea would have to take that would seriously degrade its ability to develop a nuclear arsenal.

Indeed, North Korea appears to believe that, if it definitively gave up its nuclear program in return for a mess of economic or other porridge, it might never receive it, since it would then be easier for the United States to ignore North Korea altogether. Thus, it has decided to trade concessions that do not definitely cut off its nuclear option—such as restraint on further development of its nuclear capability, permission for international inspectors to observe certain facilities or procedures, or a willingness to agree to arms control or confidence building measures—for ongoing economic and diplomatic benefits. At the same time, by postponing the "special inspections" that would shed light on its past nuclear activities, the North retains a degree of ambiguity about whether it already possesses several bombs' worth of fissile materials, or even the nuclear weapons themselves.

Thus, for the next five years or so, North Korea will receive various economic benefits from the West to make up for the support it used to receive from Russia and China. While the agreement deals specifically with oil deliveries intended to "compensate" for the loss of power from the current 5-MW reactor and the planned 50- and 250-MW reactors,[2] it seems likely that it will open the way to other benefits as well, such as Japanese investment and trade and increased remittances from the Korean population in Japan. South Korean investment may also become available.

International Atomic Energy Agency (IAEA) special inspections "before delivery of key nuclear components" of the LWR project.

[2]In fact, it is unlikely that these reactors were intended by the North Koreans to produce electric power for use elsewhere in the economy.

One important question is the extent to which the North Korean regime will allow this investment and trade to affect conditions within the country. If the DPRK were to open up its economy (e.g., if it engaged in increased trade, attracted substantial foreign investment, developed a "free trade zone" in the Chinese-Russian-Korean border area, and succeeded in establishing diplomatic relations with the United States and Japan), it might ameliorate some of its problems but exacerbate others. The economic situation would presumably improve, and the easing of its isolation might enhance the morale of the elites. On the other hand, contact with the outside world could be quite subversive in other ways, bringing home to North Koreans the economic failure of their regime (as opposed especially to the South's prosperity) and perhaps facilitating the spread of dissident ideas.

As noted, the North Korean leadership might seek to emulate the Chinese model of economic reform under authoritarian political control. (One may assume that the Chinese leadership has in fact urged them to do just this.) It is unclear, however, what the prospects for success would be: Such an opening would immediately raise the question of reunification of the Korean peninsula. What, after all, would be the justification for a separate North Korean state under those conditions? Whereas Beijing can allow its population to benefit from trade and investment relations with Hong Kong, Taiwan, and the overseas Chinese community without endangering its claim to represent Chinese sovereignty, it is not obvious that Pyongyang could do the same. Only if South Korea were willing to cooperate in forming a nominal "confederation" (that in actuality left each government fully in control of its own affairs) could the northern leadership hope to retain its legitimacy while allowing its society in fact to "converge" with that of the south.

Alternatively, the North Korean regime could attempt to preserve the major elements of the command economy; it could attempt to funnel all foreign investment through state-run firms, or, in exceptional cases, create tightly controlled and guarded free trade zones. In this way, it would attempt to derive the economic benefits of foreign contact without running the risks. In fact, we do not know how the DPRK leadership itself evaluates the relative advantages and risks of reform. The pace with which the DPRK has pursued economic reform and cooperation with foreign countries has been unhurried at best: So far, the reforms appear half-hearted.   The development zone proposed for the northeastern border with Russia and China has not progressed. All this suggests that the leadership is concerned about the possible dangers of economic reform and is uncertain how and whether to pursue it, despite vigorous Chinese encouragement. We simply are unable to evaluate how the North Korean leadership understands the balance of risks and rewards in this area, or whether there is a significant difference between the views of Kim Chong-Il and those of his father on this question.

The agreement "kicks the can down the road" with respect to the key question of North Korean nuclear intentions. That is, it is only after five years or so that the North Koreans would be obliged to take steps that would seriously degrade their nuclear option. In principle, they will have to decide, at that point, whether the future economic benefits (of the promised pair of 1,000-MW LWRs, as well as whatever other benefits are being derived from foreign trade and investment) are worth opening up the suspect sites to inspection, the dismantling of their reprocessing facilities and current reactor, and the disposal of their current stock of spent fuel (which, if reprocessed, could yield enough plutonium for up to four nuclear warheads).

Alternatively, they could attempt to renegotiate the deal at that point, depending on how they assessed their bargaining position. Many factors would be relevant here, among them, the state of U.S.-South Korean and U.S.-Japanese relations, the diplomatic stances adopted by Russia and China, the state of the conventional military balance (including the credibility of the U.S. military alliance with South Korea), and the economic health and prospects of North Korea. This last may be the most important: If North Korea were seen as able to forgo the 2,000-MW of free electric power (much more than its currently frozen nuclear program would have produced in any case), U.S. leverage, which rests primarily on our ability to obstruct delivery of key components of the LWRs, would be very much reduced.

As the recent decision by the South Korean government to permit direct trade and investment relations with North Korea suggests, the providing of the LWRs may be a relatively small part of the economic relationship between North and South by the time (the late 1990s) the North Korean government must actually take steps to constrain its nuclear program. Other Western economic interests, primarily

Japanese, may be heavily involved in North Korea by that time as well. Thus, what the relative bargaining positions of the United States and North Korea would be at that time is difficult to assess. On the one hand, North Korea would be reluctant to jeopardize the advantages it had gained through these economic contacts. On the other, North Korea's South Korean and Japanese economic partners could exert powerful pressure not to push North Korea too hard, lest the investments and opportunities for trade be lost. In particular, if the United States, without major economic interests at stake, were to attempt to take a tough bargaining stance toward North Korea, it might find that its South Korean and Japanese allies were unwilling to support it. Overall, it would seem likely that the North Korean bargaining position would be stronger at the end of the decade than it is now.

In any renegotiation of the deal, the North Koreans could seek greater economic or diplomatic benefits, could delay taking any steps that would seriously hurt their nuclear option, or both. Ultimately, however, the key question would be how the North Korean regime understands the usefulness to it of a nuclear capability. Such a nuclear capability—either declared or only suspected— could bring great benefits to the regime militarily, politically, and economically. If it in fact fears an invasion from the South, it could see nuclear weapons as a deterrent. In any case, nuclear weapons would reduce the risk to the regime of launching an invasion; if the initial thrust did not succeed in taking the entire peninsula, and if the events of 1950 seemed about to repeat themselves, the North might believe that it could always obtain an armistice on acceptable grounds by threatening nuclear escalation. (In effect, the nuclear threat would replace Chinese intervention as the regime's method of "last ditch" defense.)

Politically, a nuclear weapon capability could serve to pressure South Korea or Japan to reduce their ties to the United States. While some uncertainty on this question exists now,[3] it has barely begun to produce reactions in South Korea and Japan, or elsewhere in East Asia. How this might play itself out in the future—during the 1990s or

---

[3]The official estimate of this number is one or two weapons, although that assessment has been challenged (see Jones, 1994.)

after—is very unclear. If North Korea were to do something open and provocative—declare a nuclear capability and/or test a nuclear weapon—one would expect immediate reactions elsewhere. Alternatively, if the uncertainty simply drags on, the realization that North Korea may be a nuclear weapon state would presumably sink in more slowly and gradually and the reactions from neighboring states would be more slow in coming and perhaps harder to detect. In addition, further ballistic missile tests would bring home the danger to North Korea's neighbors.

If the regime were to be threatened with internal instability or dissidence, a nuclear weapon capability might deter outside powers from trying to take advantage of North Korea's vulnerability or perhaps even lead them to try to help stabilize the situation. Possession of nuclear weapons might be the elite's last bargaining chip in case of regime collapse.

Economically, the DPRK leadership might believe that possession of a nuclear weapon capability would further increase the willingness of the United States, Japan, and South Korea to provide foreign aid, trade, or investment advantages, to ward off the possibility that economic collapse would lead to the loss of central control of nuclear weapons and their possible acquisition and use by a faction in a civil war or a criminal group. (This is one of the arguments that has been made, for example, in favor of economic aid for Russia.)

Assuming the DPRK is serious about obtaining a nuclear weapon capability, it remains unclear whether it would stop at a small (and perhaps untested and undeclared) capability, that might be useful for deterrence; threatening escalation in extreme circumstances; and/or extracting political and economic benefits from the United States, Japan, and South Korea, or whether it would build a more substantial arsenal with greater military potential that could also produce significant revenues through sales. Of course, the North Koreans may not yet have made this choice, and their decision could be affected by the perceived costs of proceeding.

Even if, at the end of the 1990s, the North Korean regime agrees to proceed with the current agreement and takes steps to denuclearize, we would face the problem of designing an inspection regime sufficiently stringent to assure us that North Korea in fact has no nuclear weapons (or, at the least, that it has ceased its nuclear weapon development program). The DPRK's removing of the spent fuel rods from the 5-MW reactor (in the spring of 1994) without allowing the IAEA to take samples seems to have foreclosed the possibility of determining by examination of the fuel rods how much plutonium might have been diverted from that reactor. Access to the two nuclear waste sites, which under the terms of the agreement would be available in five years, might enable inspectors to make estimates of past diversion; however, if it turns out that a significant quantity of plutonium has been diverted, it might prove impossible—short of acquiring a well-placed human intelligence source—to locate it. (In theory, if the existence of previously diverted plutonium is proved by the special inspections, North Korea would be obligated to surrender it; in practice, however, it is much more likely that North Korea would simply dispute the IAEA inspectors' conclusions and deny that the material existed.) In this case, assuring ourselves of the nonexistence of any nuclear weapons would be very difficult, if not impossible.

Assuring ourselves that the nuclear program has completely ceased depends on our ability to be sure that we know the location of all relevant facilities. The experience of post–Gulf War Iraq suggests that this may be much more difficult than we might think. At a minimum, we would need North Korean agreement to allow the IAEA to conduct "special inspections" of *any* nondeclared sites of which we might become aware through technical or other types of intelligence collection; it is not clear whether the current agreement provides for such access or whether, if it did, such access would be granted in the post-2000 period.

This review of the situation, both as it exists now and as it is likely to exist into the next decade, suggests that we are likely to have to figure out how best to live with at least the uncertainty about whether North Korea possesses a nuclear capability—during the next five or so years, during which North Korea will be exempt from IAEA special inspections, as well as in the subsequent period. The current agreement, however, assuming that there are no North Korean facilities of which we are ignorant, enables us to limit the range of uncertainty to the number of weapons that could be fabricated from plutonium diverted before 1994.

In any case, general recognition of the possible existence of a North Korean nuclear arsenal will create military and political requirements and consequences. Militarily, we would want to bolster South Korean and Japanese ballistic missile defenses (given North Korea's ballistic missile program), as well as the conventional defense of South Korea. (As noted above, North Korea may see its nuclear weapon capability as a means of reducing the risk to it of conventional aggression.) While we would continue to oppose any Japanese nuclear capability, we would have to expect that Japan would pursue a "virtual" capability, i.e., the ability to acquire nuclear weapons relatively unobservably and in a short period of time given the political decision to do so.

It is at least possible that South Korea would want to acquire its own actual or "virtual" nuclear capability as well. In the 1970s, the military regime's nuclear ambitions were successfully blocked by U.S. pressure on it and on its suppliers, but it is unclear whether today's civilian government would wish to resume such a program. We would have to decide whether a nuclear capability would be stabilizing and, if so, whether the perceived benefits would outweigh general nonproliferation goals. We would also have to take into consideration the effects on South Korea's internal political situation; if the democratically elected civilian government were to suffer the defeat inherent in the North's acquisition of nuclear weapons and then bow to U.S. pressure not to develop its own, would that affect its prestige in ways that might have serious consequences?

Political steps to bolster South Korean and Japanese confidence in the United States would be necessary, especially if it were necessary for the United States to oppose actively these countries' pursuing their own nuclear capability. Alternatively, South Korea might become more intent on securing a "soft landing" for North Korea (i.e., avoiding any instability or collapse of the regime) and hence insistent on continuing its policy of enhanced contacts, even if the United States wished, in the late 1990s, to exert more pressure on North Korea to force it to live up to its obligations with respect to the dismantling of its nuclear program. The United States would have to decide whether to support such a policy, which might be harmful for global nonproliferation policy, since it would look like a reward to North Korea for its violation of the Nuclear Non-Proliferation Treaty.

With respect to Japan, it might prove difficult to measure the extent to which confidence in the United States was damaged or the danger that Japan might abandon its policy of close reliance on the United States in national security matters, until it took some dramatic step. Closer political-military consultation with the Japanese government would be necessary to make sure that we did not suddenly discover that Japan was determined to pursue an independent course.

As provided for in current strategy documents, the U.S. armed forces, during this entire period, would, together with the South Korean forces, play a major role in deterring and, if necessary, defeating a North Korean attack on the South. Given the recent agreement, it is unlikely that the Team Spirit exercises will be held in the future, thereby requiring that other mechanisms for maintaining readiness be devised. Political pressures, both in the United States and South Korea, for reducing conventional forces are likely; however, since the agreement does not limit North Korean conventional forces or their forward deployment near the demilitarized zone, the military requirements for deterrence remain unaffected. In particular, North Korean ballistic missile development is unconstrained by the agreement, implying that the requirement for ballistic missile defense remains unchanged as well.

LONG-TERM DEVELOPMENTS AND POSSIBILITIES

The Korean peninsula presents a strange paradox: Nobody knows what might happen this year or next, but (almost) everybody agrees on how things will look in ten or twenty years. The reason for this is the peculiar character of the North Korean regime: belligerent, armed to the teeth, impulsive, and, consequently, quite unpredictable at present but, according to the conventional wisdom, doomed in the long run by economic decline and political, ideological, and diplomatic isolation.

From 1989 to 1991 and in light of the events of those years, it seemed obvious that the remaining communist regimes could not last long. Now, perhaps, things are less clear. The reforming countries, in particular, China and, to a lesser extent, Vietnam, seem to have good economic prospects that may enable their authoritarian leaderships to survive, although economic development will itself bring great stresses. Nevertheless, the electoral success of communist and excommunist parties in Eastern Europe and the former Soviet Union suggests that, if the communist parties of China and Vietnam can manage economic reform, they may be able to remain in power.

At the same time, the nonreforming regimes, most notably Cuba and North Korea, which several years ago were all but written off, have survived as well, despite great economic difficulties. It may be that the key here is the longevity of the regimes' founders, Fidel Castro and Kim Il-Sung, and that these regimes will not long survive their deaths. As noted above, Kim Chong-Il lacks his father's prestige among the leadership elite and the military leadership, in particular among whom Kim Il-Sung had been a dominating figure for decades. The generally calm nature of the current succession period may be superficial and misleading.

On the other hand, both regimes have shown remarkable resiliency in the face of the ideological shock of the collapse of communism elsewhere and the economic problems that resulted therefrom due to the loss of subsidies and opportunities for barter on favorable terms. If Kim Chong-Il can consolidate power rapidly, he will inherit a strong repressive apparatus. Thus, however unlikely, it does not seem impossible that—assuming he is physically up to the challenge—he will be able to continue to rule North Korea in the same manner as his father. In particular, much depends on whether the North Korean regime will be able to reap the economic benefits of increased contact with South Korea and Japan and the diplomatic benefits of its newfound international legitimacy[4] without jeopardizing its political-ideological foundations; generally speaking, the nuclear agreement reduces the odds of a collapse of the North Korean regime.

If, for the purposes of discussion, we accept the proposition that the North Korean regime is doomed in the long run, the major question becomes the manner of its death. Having observed the difficulties that (much richer) West Germany is having in absorbing the excommunist East, the South Koreans seem to have decided that they would prefer a "soft landing," i.e., a long period during which the

---

[4]For example, on November 7, 1994, Secretary of State Warren Christopher announced that the United States would support North Korea's entry into the forum for Asia Pacific Economic Cooperation.

North Korean economy develops (presumably with capital from, and under the influence of, South Korea) toward the South's level. When unification finally comes, under this scenario, the costs and difficulties of adjustment will be much less than if it occurred suddenly. Of course, this option presupposes that nothing untoward will happen in the meantime; while the recent agreement silences, as a political matter, the argument that the continued existence of the northern regime poses real military dangers, the eruption of a new political-military crisis cannot be ruled out.

The South Koreans' fear of a North Korean "implosion," which would force them suddenly to take responsibility for the welfare of the North's citizens, seems to have played a role in their reluctant attitude toward the imposition of serious sanctions (e.g., on financial remittances from Japan and on oil imports).[5] This is in addition to the calculation that, if the northern regime is doomed anyway, there is no point in risking a war over the nuclear issue.

Despite this preference on the part of the South, a relatively rapid collapse in the North cannot be ruled out. Despite the surface calm, there might still be a power struggle before Kim Chong-Il consolidates his power, or if he proves unable to hold on to it. Any such struggle at the center in a regime as rigid and totalitarian as that of North Korea could have explosive consequences. We simply do not know enough about the regime's internal workings to speculate on how it might evolve, whether the participants would try to mobilize other groups or constituencies in the society and, if so, which ones. One would expect that the armed forces and the secret police would be key constituencies to which contenders for power would appeal. If nuclear weapons have been developed, control of them might be a key factor in the power struggle.

---

[5]It is worth noting that the South Korean view of the costs of unification may not be accurate. Mainly for domestic political reasons, Chancellor Kohl of Germany chose to give the East Germans a very favorable exchange rate for their currency, thereby creating inflationary pressures and raising East German wages to unrealistically high levels. To some extent, Kohl was also driven by the fear of a mass migration to the West where, even without jobs, East Germans would enjoy the benefits of one of the world's most generous welfare states. It would thus appear probable that South Korea (whose leadership might be under less domestic political pressure and which is not a welfare state) could avoid these difficulties if it wanted to.

The North Korean regime could also collapse in some other manner. If we look at the experience of the East European communist states, the examples of East Germany and Romania appear to be the most relevant, the former because it, like North Korea, was the communist part of a larger nation and the latter because, within the communist world, only it approached North Korea in the extent to which government had become a family business. The former example suggests that a gradual reform effort could create a groundswell for immediate unification that would be unstoppable. The second suggests that obscure fissures within the ruling elite could, given the right public spark, quickly burst into violence and revolution.

As noted, most observers believe that Korea will be unified in ten or twenty years time, essentially under the current southern government. Once the costs and difficulties of the unification process had been surmounted, a united Korea would have the potential to become a major industrial country (with a population of about 70 million) and a key factor in the Northeast Asian security environment.

Initially, at least, a united Korea that inherited the armed forces of both North and South would, at least in order-of-battle terms, be stronger than Japan. One would expect that the ground component of the united force would be decreased rather rapidly in the absence of a major ground threat. On the other hand, a united Korea might maintain relatively strong naval and air forces. It would also inherit the North's nuclear program, which it might be hesitant to dismantle.

Korea, a "minnow among the whales," borders or is close to the three major powers of Northeast Asia: China, Russia, and Japan. Of the three, Korea is likely, for historical reasons, to regard Japan with the greatest suspicion and mistrust. As the recent controversy concerning the "comfort women" indicates, the animosity engendered by Japan's decades-long occupation of Korea has not been resolved. In addition, of the three countries, Korea is most likely to see itself as an economic competitor of Japan.

China, on the other hand, might enjoy a much closer relationship with a united Korea. Aside from a common wariness about Japan, the economies of the two countries are likely to be relatively complementary, as the currently rapid growth of Chinese trade with South Korea would indicate. Korea is relatively resource-poor and would be a potential importer of Chinese oil and other raw materials; in return, it could become a source of modern manufacturing expertise and machine tools, as well as capital, for China. Similarly, Korea might seek a close economic relationship with Russia and Central Asia, providing relatively inexpensive (as compared to, say, Japan) high technology products in return for energy and other raw materials.

It is not clear what attitude a united Korea would take toward the U.S. military presence in the region. Given Korea's "minnow among the whales" situation, maintaining a strong security link with the United States would seem to be simple prudence. At the same time, nationalist sentiment might require the removal of U.S. ground troops once the immediate cause for their deployment—the threat from the North—was obviated. In addition, China might be much more sensitive to U.S. troop deployments in Korea once the North Korean "buffer" no longer existed; Korean sensitivities to Chinese views would also play a role in the formation of its policy toward a possible U.S. presence. New kinds of access arrangements that did not involve permanent basing, as well as combined exercises and other types of military cooperation, would have to be worked out.

It is at least possible that the unification of Korea would be a major shock to Japanese opinion, causing it to take foreign policy issues much more seriously than it has in the past. This shock would be exacerbated by the fact that the immediate, "emotional" reaction to Korean unification in the United States and Japan could be quite different. Public opinion in the United States would likely welcome unification, both because it would remove one possible cause of U.S. involvement in a war and because it would right a historic injustice done to the Korean people. Japanese opinion, on the other hand, would likely take a dim view of the sudden emergence of a heavily armed unified Korea, with a population of 70 million, on Japan's doorstep. The result could be an impetus toward a more independent Japanese security policy than we have seen in the post–World War II period.

In the absence of closer Korean-Japanese relations, the United States, especially the U.S. armed forces, could provide a vital communication link between the two nations' militaries. In the immediate postunification period, Japan might be very sensitive to the pace at which Korea's military forces were being reduced and to the manner in which they are being restructured; Japanese suspicions might be aroused if the ground force reductions occurred slowly or if the emphasis shifted abruptly to air and naval forces. The U.S. armed forces would have a major role to play in interpreting each nation's military strategy to the other and in preventing misunderstandings from arising. More generally, a U.S. military presence in the region would likely be reassuring to both sides, and hence an important force for stability.

REFERENCE

Jones, Gregory S., *How Many Nuclear Weapons Does North Korea Have: Will We Ever Know?* Santa Monica, Calif.: RAND, PM-253-AF, June 1994.

Chapter Eight

CHINA

Michael D. Swaine

INTRODUCTION AND BACKGROUND

China is undergoing extremely rapid and revolutionary change along every dimension of national power: economic, political, military, technological, and social. A systematic program of market-led economic reform and opening to the outside inaugurated in the late 70s has produced growth rates of nearly 10 percent per annum since the 80s. This explosion in growth has resulted in major increases in living standards for most of the population, a loosening of political controls over society (and rising expectations of further change), strong and expanding economic and diplomatic linkages to nearby Asian countries, and a determined effort to construct a more modern and comprehensive military establishment. However, it has also created severe disparities in income, periods of high inflation, increasing numbers of displaced and unemployed urban and rural workers, and growing corruption at every level of the polity and society. To complicate matters further, China is also facing an unprecedented leadership transition to a new generation of elites. Although largely united in their commitment to the maintenance of economic growth and the enhancement of national wealth and power, these new leaders possess less authority and arguably less vision than their predecessors.

These events pose major implications for the future. Indeed, China today arguably constitutes the most critical, and arguably the least understood, variable influencing the future Asian security environment and the possible use of U.S. military forces in that region. If current growth trends continue well into the next century and if most of the problems mentioned above are overcome, China could emerge as the dominant military and economic power in Asia, capable of projecting air, land, and naval forces far from its borders while serving as a key engine of economic growth for many nearby states. Such capabilities could embolden Beijing to resort to coercive diplomacy or direct military action in an attempt to resolve in its favor various outstanding territorial claims or to press other vital issues affecting the future economic and security environment of the region.

Such a troubling development could eventually reorder the regional security environment in decidedly adverse directions. For example, a Sinocentric system might emerge in which most Asian capitals generally defer to Beijing's interests and needs and ignore or downplay those of the United States and its allies. Alternatively, an increasingly assertive China could produce a highly volatile and unpredictable regional environment, marked by escalating confrontations with other existing or such rising regional powers as Japan, the Association of Southeast Asian Nations (ASEAN) states, and perhaps India.

Equally negative outcomes for Asia would also likely emerge from a reversal or wholesale collapse of Beijing's experiment in combining political authoritarianism with liberalizing market-led reform. National fragmentation, breakdown, and/or complete chaos could result, leading to severe economic decline and a loss of government control over the population and over China's national borders. Such developments would almost certainly generate massive refugee flows and send economic shock waves across the region, producing major crises for neighboring countries.

Other far less turbulent outcomes are also possible, however, including the emergence of a more stable and cooperative China with limited military capabilities and strong economic interdependencies with other countries in the region. This outcome would probably require the creation of a new leadership coalition in Beijing, including more open-minded and progressive individuals from both the center and the provinces and perhaps even some members of the emerging semiprivate business community. Such a coalition might arise in the context of the gradual democratization of the Chinese political system, preceded by further economic reform and relatively high growth rates.

This broad range of possibilities and, in particular, the considerable potential for adverse developments demands a careful assessment of the forces driving change in China today. This chapter examines those changes under way in China that pose the greatest implications for the future use of U.S. military power. It especially stresses the determinants and characteristics of China's changing security and defense policies and the related program of military modernization, particularly those elements affecting the future evolution of Chinese air and naval forces.[1] The final section of this chapter specifically addresses some of the possible implications such developments pose for U.S. airpower.

CHANGES IN CHINA'S STRATEGIC ENVIRONMENT

China's strategic environment has undergone enormous changes since the late seventies, largely as a result of the interaction of three key developments.

A Changed Security Environment

The collapse of the former Soviet Union and the explosion of economic growth throughout most of Asia have produced a far less threatening yet arguably more complicated and uncertain security environment for China. From Beijing's perspective, five central features define this environment:

- A powerful and potentially threatening United States, increasingly at odds with China over a host of issues from human rights to arms sales and bilateral trade, still the dominant military power in Asia, yet also indispensable as an effective counterweight to Japan and an essential market for Chinese exports

---

[1]Much of this discussion is taken directly from several of my recent works, including Swaine (1994b) and (1994c).

- An economically powerful and increasingly independent Japan, less closely tied to the United States, with expanding trade and investment links to China and other nearby Asian areas, high absolute levels of military spending, and a growing capability to develop offensive conventional (and possibly nuclear) weapons

- A more militarily capable and economically emergent India, with growing maritime interests, increased attention to Southeast Asia (historically the focus of Sino-Indian geopolitical competition), and decades of rivalry and sporadic border conflict with China

- A host of rising second- and third-tier Asian powers (including South Korea, most of the ASEAN countries, and Taiwan), with rapid growth rates and expanding foreign trade and investment links, greater attention to external (especially maritime) strategic interests, and increasing air and naval capabilities[2]

- The emergence of relatively unstable Islamic states on China's Central Asian borders, economically undeveloped and potentially threatening to those Chinese regions containing large Muslim minorities, such as Xinjiang Province.[3]

The United States is viewed with particular suspicion as the only remaining military superpower, seeking to dominate regional and global events but increasingly constrained by both growing competition with emerging major powers such as Germany and Japan and its own internal economic and political problems. Virtually all Chinese strategists and officials believe that the United States will continue to exert a decisive influence over China's future security environment. Unfortunately, most of these individuals also believe that future U.S. geopolitical interests will require a relatively weak and divided China. In fact, many in China's policy community argue that the United

---

[2]Taiwan clearly constitutes a special case among these powers, since Beijing regards it as a part of China.

[3]This listing is not intended to suggest that the Chinese leadership is unconcerned about the potential threat posed by a resurgent Russia. However, in the author's view, such a possibility is usually relegated to the distant future by most Chinese strategists. Moreover, some strategists believe that Sino-Russian interests might actually converge in the future, if Moscow's experiment with democracy fails and Russia reestablishes a socialist or authoritarian government.

States will increasingly strive (or be compelled) to prevent China's full emergence as a major economic and military power.[4]   At the same time, however, the United States is also seen as indispensable to Chinese development, as China's major trading market, a key source of technology and knowledge, and the educator of thousands of Chinese engineers and scientists.   Moreover, a continued, yet reduced, U.S. military presence in Asia is seen by many Chinese strategists and officials as beneficial to the maintenance of general stability in the region, at least through the end of the century.[5]

Similar ambiguities confront China's assessment of Japan.   On the one hand, Tokyo is viewed by many Chinese strategists and officials as Beijing's primary potential geopolitical foe (and possible regional economic competitor) over the long term.   Despite Tokyo's protestations to the contrary, many Chinese defense planners in particular remain concerned that Japan will eventually translate its enormous and growing economic power in Asia into significant political and perhaps military influence, thus posing a major challenge to China's strategic position in the region.   On the other hand, Japan is also seen as a key source of economic, financial, and technological assistance.

---

[4]We should note, however, that important differences exist among Chinese strategists and political leaders over the extent to which Sino-U.S. tension or conflict is viewed as inevitable over the long term.   More broadly, Chinese observers also differ over how long U.S. global dominance in the post–Cold War era will last and what form it will take.   Some Chinese believe that the U.S. desire to retain dominance will lead to greater cooperation with Germany and Japan, while others argue that the United States will be driven by intensifying economic competition to contend with both countries and eventually seek dominance over them.   This could result in increasing global conflict.

[5]This last point is made by Garrett and Glaser (1994, p. 14), which was based on information derived through extensive interviews conducted in China during summer 1993.   The authors argue that most Chinese civilian and military strategists are convinced that a dramatic and rapid reduction of the U.S. presence in the region would prove politically and strategically destabilizing, despite official rhetoric concerning opposition to foreign military bases and Beijing's suspicions regarding U.S. intentions.   The Chinese are particularly concerned that such actions would result in an expansion of Japan's military role and a possible regionwide arms race.   However, we should note that considerable controversy within the Chinese security policy community almost certainly exists over the duration and extent of the U.S. military presence in Asia.   For example, many military analysts insist that Chinese interests over the medium to long term (5–15 years) would be best served by a major (if not total) U.S. military withdrawal from the region, combined with a continued political and economic presence.

Moreover, good relations with Japan provide China with increased potential political leverage in dealing with Washington.[6]

Despite recent significant improvements in the Sino-Indian relationship, Beijing's concerns about New Delhi have increased in recent years. The major causes of such concerns include growing Indian naval capabilities, a more flexible and cooperative Indian diplomatic and military approach to many ASEAN states, and the overall modernization of New Delhi's military forces, including improvements in various power-projection components. These developments present the potential for increasing Sino-Indian politico-military competition (and perhaps friction) over Southeast Asia over the long term. In addition, India's movement toward a free-market economy focused on expanding foreign trade and investment ties suggests a basis for increasing economic competition with China in Asia.

The growing economic and military prowess of critical second- and third-tier Asian states, along with their increasing outward orientation, greatly complicates China's long-term strategic ambitions, especially regarding territorial claims in such areas as the Spratly Islands in the South China Sea and, of course, toward Taiwan. In both areas, China confronts the acquisition of significant power-projection capabilities by potential or actual adversaries. Moreover, on the diplomatic level, the Chinese are increasingly concerned that their efforts at military modernization and economic development will lead to the emergence of a tacit anti-China coalition among many ASEAN countries. In this context, the inclusion of Vietnam in ASEAN (along with the continued improvement of Hanoi's relations

---

[6]Because of such ambiguities, Chinese strategists and officials differ over the exact nature and extent of the threat posed by Japan. Some believe that Japan will inevitably view China as its major economic rival and security threat and thus seek to contain China or prevent its emergence as a major power, either in alliance with the United States or independently, if the U.S.-Japan alliance collapses. The latter development would likely lead to Japan's acquisition of significant offensive conventional and even nuclear capabilities. Still other Chinese strategists believe that Japan will avoid efforts to contain or restrain China, and resort instead to a more sophisticated strategy designed to force Washington to confront Beijing, while Tokyo maintains positive relations with both powers. A third (minority) viewpoint argues that Japan could eventually ally with China to lead a Pan-Asianist, anti-U.S. political and economic bloc. For further details, see Swaine (1995), pp. 52–53, 86–87.

with Taiwan and the United States) is viewed with particular alarm.[7] Similarly, Taiwan's improved international standing[8] and rapid movements toward democratization have increased the island's actual and potential diplomatic leverage in dealing with China, thus adding to Beijing's anxieties.[9]

On the other hand, the emergence of second- and third-tier Asian countries also presents certain potential economic and diplomatic advantages to China. In particular, such a development offers new export markets, more robust sources of manufactured inputs and much-needed capital, and, in the case of South Korea, more influential potential strategic partners against possible future foes, such as Japan and a resurgent Russia. Moreover, close diplomatic and economic links between China and several Southeast Asian states would likely increase Chinese leverage in handling an emergent India over the long run, while also helping to restrain future possible U.S. and/or Japanese efforts to contain China.

---

[7]One reason some Chinese strategists oppose a major U.S. withdrawal from Asia is that such an action would greatly encourage the emergence of such a (tacit) coalition among ASEAN countries. Fear of such a development is also one reason behind China's opposition to a more structured multilateral approach to security issues in the Asia-Pacific region. More on the latter point below.

[8]As a result of President Lee Teng-hui's multifaceted strategy of "flexible diplomacy," Taiwan is again participating in major international organizations, such as the Asian Development Bank; has improved links with many countries, including the former communist states of Eastern Europe, as well as Russia and several of the former Russian republics; has established formal diplomatic ties with several nations; and in general enjoys expanding economic relations with an increasing number of countries, including Vietnam and many members of ASEAN. For further details, see Wu (1994, pp. 46–54).

[9]The level of Beijing's concern over Taiwan's increasingly successful diplomatic forays increased dramatically as a result of President Clinton's decision, under U.S. congressional pressure, to grant Taiwan President Lee Teng-hui a visa to visit Cornell University in June 1995. This incident convinced many Chinese leaders that Lee Teng-hui is pursuing a diplomatic strategy aimed at the attainment of Taiwan's *de jure* independence from the mainland, despite his public statements to the contrary. Such a fear has arguably strengthened Beijing's willingness to use coercive diplomacy, if not outright force, against Taiwan, as suggested by the markedly more threatening nature of various Chinese air, land, and naval exercises that followed Lee's U.S. visit. These included a series of unprecedented firings of land-based mobile ballistic missiles and naval missiles into waters near Taiwan. These developments have dramatically increased the level of military tension across the Taiwan Strait, while significantly damaging U.S.-China relations. As a result, the "Taiwan problem" has arguably become Beijing's primary security concern at present.

Finally, the existence of unstable ethnic regimes on China's Central Asian borders arguably poses the most direct and unambiguous external threat to domestic order within China. This is especially the case given the history of severe ethnic unrest in the northwest and the increasing amount of contact between Muslim minorities on both sides that has resulted from the explosion in cross-border trade under the economic reforms of the 80s and early 90s. As a result, Beijing is increasingly sensitive to any signs of adverse influence along its western border originating from the former Soviet republics of Central Asia.

Overall, Chinese leaders and strategists believe that such developments have produced a more complex, multipolar strategic environment and a much greater regional emphasis on economic and technological (as opposed to purely military) means for advancing state interests and ensuring security. For China, these changes clearly present a mixture of opportunities and challenges. On the one hand, they have strengthened confidence in Beijing's ability to avoid major conflicts and influence external events in Asia, given the absence of a single, sizable external threat, the greater opportunities for diplomatic maneuver afforded by increasing multipolarity, and the likely benefits resulting from a common regional stress on economic development (more on the last point below). On the other hand, such changes also present a wide range of instabilities and uncertainties that serve to temper Chinese confidence, especially over the medium to long term.[10] From the Chinese perspective, the above features will likely produce an increasing number of small conflicts and military-related disputes (termed "local wars") around China's periphery, many linked to conflicts over territory and perhaps involving regional arms races and weapon proliferation issues.

---

[10]It must be added that China's sense of vulnerability to external pressures is greatly enhanced by the insecurities and anxieties created by its rapidly changing domestic environment, marked by the passing of the original revolutionary generation that has provided leadership cohesion and continuity since the founding of the People's Republic, the declining financial and administrative capabilities of the central government, the increasing economic (and potential political) influence of various new social and economic classes created by the reforms, growing economic and political corruption at all levels of society, and the growth of regional power centers, many linked closely to foreign markets. In traditional Chinese thought, such internal regime weakness and social division is seen to provide opportunities for foreign aggression. For details on all these points, see Swaine (1995).

Such instabilities will also likely generate various nonmilitary problems, including those derived from intensified economic competition with many countries in Asia, especially the United States and Japan, and an array of diplomatic challenges.[11]

A Changed Economic Environment

The second major development influencing Beijing's foreign and security policies is the emergence of a booming reform-based, market-led Chinese economy with expanding oceanic links and growing external dependencies. The former closed Soviet-style planned economic system of the Maoist period, marked by collectivized agriculture and heavy industrial production through huge state enterprises, is being gradually replaced by a largely decentralized, open, and market-driven economy increasingly keyed to the manufacture of light consumer goods for foreign and domestic markets. This transformation has brought about revolutionary changes in Chinese production levels, patterns and volumes of manufacturing and trade, personal income levels, government revenues, and foreign exchange earnings.[12]

Specifically, the Chinese economy has been expanding at an annual average rate of more than 9 percent since 1980. Such rapid growth has produced major increases in living standards across most of the country, especially in China's increasingly cosmopolitan eastern and southern coastal areas. Much of this economic and social development is driven by rapidly expanding economic links with the outside, in the form of huge increases in external trade and major inputs of foreign finance and technology from many states and territories around China's periphery (most notably Taiwan, Hong Kong, Japan, South Korea, Thailand, Malaysia, and Singapore), as well as with the United States and Western Europe. These developments are increasingly strengthening the bonds between China's economic expansion

---

[11]Of course, such military and nonmilitary problems are closely interrelated. For example, increasing economic development and competition in Asia could lead to military confrontations or conflicts over scarce natural resources. Conversely, in the Chinese view, "local wars" could be constrained or modified by diplomatic, political, and economic pressure, and a solution could in some cases be attained through negotiation and compromise.

[12]This paragraph is taken from Swaine (1995), p. 57.

and Asian patterns of growth and hence of the trade and investment decisions made by political and economic elites throughout the region. As a result, China is fast becoming a major overseas trading nation with a strong interest in continued regional and global development and an increasing dependence on foreign markets, investment, technology, and management knowledge.

These changes in China's domestic and external economic environments have created the wherewithal for a more robust level of military spending and have engendered a growing Chinese interest in the defense of trade routes and the expansion of access to and influence in nearby Asian waters. They have also presented Beijing with a new source of potential diplomatic leverage in Asia and beyond. At the same time, however, China's increasing level of dependence on foreign economic relations also serves to restrain Beijing from weakening or reversing the open-door policy. High economic growth rates through continued reform and contacts with the outside are now essential to the maintenance of social order in China, given rising expectations of a better life among the populace and the obvious delegitimization of communist party doctrine and authority that has taken place in recent decades. Moreover, without continued economic development, the Chinese leadership will likely prove unable to resolve many of the major problems plaguing society, such as severe income inequality, insufficient government income, and rampant corruption.

A Changed Military Environment

The third key change in China's strategic environment is considerably narrower in scope than the previous two, relating more directly to defense capabilities: the emergence of the high-tech battlefield. This revolutionary change in warfare was exemplified by the performance of U.S. forces during the Gulf War of 1991, during which a wide array of highly sophisticated American weaponry and other military capabilities was deployed to defeat, in short order, a very large Soviet- and Chinese-equipped Iraqi force. The U.S. capabilities that most stunned the Chinese leadership included precision-guided munitions; stealth technology; the high volume of aircraft sorties; airborne command and control systems; satellite-based targeting; intelligence gathering; early warning and surveillance systems;

coordinated large-scale naval, air, and land attacks; multifaceted night warfare capabilities; and the effective use of rapid deployment and special commando units (Shambaugh, 1994, pp. 3–15).

The generally low quality of Chinese forces and the obsolescence of the military's combat tactics and strategy had been previously indicated by the very poor showing of PLA units during the Sino-Vietnam border conflict of 1979 and 1980. The hard lessons learned from that experience had already convinced China's leadership of the need for significant military reform. However, the Gulf War confirmed in an even more convincing manner the argument that Chinese defense strategists had been making to the Chinese military high command for many years, i.e., that China's existing force structure, operational doctrine, training, deployments, and C^3I capabilities stood little chance against a highly mobile, well-organized, and coordinated land, sea, and air force armed with a wide variety of precision-guided weaponry and fully capable of fighting under almost any conditions, day and night.

Operation Desert Storm confirmed the obsolescence of the Maoist notion of People's War, centered largely upon a protracted war of attrition against a massive conventional invasion, conducted by large numbers of slow-moving infantry and armor-led forces, backed by reserve and militia units engaged in guerrilla warfare. This doctrine relied essentially on the use of World War II–era ground warfare tactics involving massive numbers of foot soldiers (i.e., "the human factor"), largely armed with light weapons and deployed in mobile combat along a fluid front.[13] In place of these features, weapons, technology, and systems for the rapid, coordinated deployment of smaller yet more sophisticated air, land, and naval forces are now viewed by most Chinese strategists as ". . . the decisive elements in modern warfare." (Shambaugh, 1994, p. 13.) In addition, in terms of

---

[13]It should be noted, however, that the concept of People's War had already undergone some modification in the early eighties, partly in reaction to the lessons of the Sino-Vietnam border conflict. These changes led to the notion of "People's War Under Modern Conditions." This doctrine placed a greater stress, for example, on positional warfare and the defense of cities, and on initial battles close to the Chinese border, rather than "drawing the enemy in deep," a basic tenet of People's War. But basic assumptions about the scope and nature of warfare (e.g., global war involving a massive invasion of China) and the structure of forces remained essentially the same. For details, see Nan (forthcoming).

the scale of conflict, the greater destructive power and enormous cost of high-tech conventional weaponry reinforced, in the Chinese view, the above geopolitical reasons for the increased likelihood of limited or local war over global war. Such modern weaponry created strong incentives to deploy fewer troops, to seek a quicker resolution of the conflict, and to contain the size of the battlefield.

In sum, the Gulf War (1) showed China's top civilian and military leaders just how far behind the Chinese military had fallen, in both technological and doctrinal terms, and how much effort was needed to "catch up" with advanced industrial states; (2) reaffirmed the correctness of the general conclusions drawn from China's changed security environment regarding the type of intense, short, limited-scope conflicts that will likely occur in the future; and (3) ended the acceptance of People's War as a credible strategy of defense for China in the post–Cold War era.[14]

CHINESE SECURITY POLICY

The above changes in China's strategic environment led to the emergence, in the 80s, of three major features that together are the foundations of China's current post–Cold War security and defense policies:

- A notion of "comprehensive national strength" that posits the primary strategic importance of continued civilian economic, technological, and social development and the secondary importance of traditional military-related goals

- A diplomatic approach to great-power relations that stresses the search for greater strategic leverage through a more complex version of realpolitik, balance-of-power politics, including the use of economic incentives

---

[14]However, this does not mean that Chinese strategists and officials have unanimously accepted the need to immediately and drastically reduce the size of China's ground forces. (This point is discussed below.) Neither does it mean that the doctrine of People's War has been entirely discarded by low-level field units. According to knowledgeable observers, the above more modern concepts of warfare have yet to percolate down to the most basic units.

- A transformed force structure and defense doctrine centered on the concepts of local war, active peripheral defense, and limited air and naval power projection, reflecting both continental and maritime strategic concerns.

Comprehensive National Strength and a Priority on Economic Development

The Chinese concept of "comprehensive national strength," first evident in the 80s, reflects Beijing's awareness of the need to develop all the dimensions of national power (i.e., economic, political, military, technological, and social) to attain the status of a great power in the post–Cold War world. Equally important, this concept also contains a clear prioritization of these dimensions of power, reflecting the major characteristics of China's altered security and economic environments. In particular, military reform and modernization have been made subordinate to and dependent upon civilian economic growth, given the pivotal importance of economic and technological development to social stability, defense modernization, and the successful pursuit of Beijing's diplomatic strategy, noted above.[15]

As a result of this new prioritization, China's primary foreign policy objective has become the maintenance of a placid regional and global environment conducive to the successful implementation of domestic reform and the creation of a strong, modern economy. This objective is to be achieved through continuing the open door policy in external economic relations; expanding economic and diplomatic ties with all Asian states; lowering the probability of armed conflict; and maintaining reasonably good relations with the United States, Europe, and Russia, as noted above. In this calculus, traditional foreign policy objectives directly associated with the development or use of military force, such as the defense of national sovereignty and unity and the attainment of big-power status, are thus relegated to secondary positions.[16]

---

[15]Much of this discussion is taken from Swaine (1995).

[16]During the prereform period, civilian economic development was viewed as an essentially separate, domestic issue, associated with internal aspects of mass mobilization, collectivization, and social transformation. Thus, in the context of the Cold

However, both of these traditional foreign policy objectives are deeply rooted in growing nationalist attitudes, primarily associated with the military and conservative civilian elites, and thus remain extremely important in Beijing's strategic outlook.[17] Hence, while some Chinese leaders and strategists (and many ordinary citizens) may value civilian economic development primarily for the domestic prosperity and stability it brings, many others undoubtedly view it as largely an instrument for the eventual attainment of China's great power ambitions toward the Asia-Pacific and beyond. This suggests that the Beijing leadership might eventually reverse the above prioritization, once it believes China has attained a relatively advanced level of economic and military development, and pursue more assertive nationalist objectives with less regard for the maintenance of a placid and stable regional or global environment. Yet such a reprioritization would almost certainly generate significant controversy within the Chinese leadership and strategic community, given the obvious and enormous risks entailed (more on this point in the final section of this chapter).

Reliance on the Modalities of Realpolitik, Balance-of-Power Politics

From Beijing's viewpoint, the emergence of a more complex, multipolar security environment in Asia provides China with renewed opportunities to attain the above major foreign policy objectives and to deal with the above uncertainties and concerns. As in the past, China's diplomatic approach remains keyed to the search for strategic leverage and greater independence of action through the exploitation of rivalries and the balancing and manipulation of relations among both major and emerging powers. Among the former countries, such efforts are often designed to weaken, break up, or

---

War and growing tensions with the Soviet Union, defense of national sovereignty and the attainment of big-power status largely dominated Chinese strategic thinking.

[17]For a more detailed discussion of the increasing influence of conservative nationalist viewpoints upon Chinese foreign and defense policies, see Swaine (1995), especially pp. 31–34 and 45–46.

prevent the emergence of a dominant power or an alignment of powers opposed to China.[18]

Many Chinese believe that the Soviet–United States–China strategic triangle of the Cold War era has been replaced by a new and more complex geopolitical and geoeconomic strategic triangle between China, Japan, and the United States, albeit with additional complications presented by the emergence of secondary players, such as India, South Korea, Taiwan, and many of the ASEAN states.

Chinese behavior in such an environment employs a flexible, sometimes conciliatory and often expedient diplomatic approach. Central elements include the following:

- The search for closer diplomatic relations with potential economic and political rivals of the United States, such as Japan, Germany, or India

- The development of common interests with most Third World (and especially Asian) states, to raise China's global stature and increase Beijing's bargaining leverage with the United States and Japan, especially on important economic and political issues

- Increased, albeit highly limited, support for multilateral approaches to various Asian security issues, primarily intended to allay fears concerning China's future intentions toward the region while minimizing constraints on Chinese behavior

---

[18]While reflecting the specific characteristics of the post–Cold War era, it should be pointed out that reliance on such realpolitik practices originates from deeply held Chinese historical and cultural attitudes toward international relations, centered upon two fundamental beliefs:

- That each nation acts always and entirely out of independent self-interest, seeking to maximize the attributes of national power to the extent possible, while maneuvering to balance, neutralize, or complicate the efforts of real or potential adversaries

- That the world is in constant flux and disequilibrium, with some nations in the ascent and others in the descent; hence, China must always seek to maximize its independence and maintain diplomatic flexibility, avoiding close alliances or relations of interdependence.

For further details, see Shambaugh (1992, 1993). For additional insights into Chinese security thinking, see Wang (1992).

- Support for the full resumption of official political and military dialogues and exchanges with the United States and its allies, combined with limited concessions on major U.S. concerns, such as human rights, arms sales, and trade

- Maintenance of positive relations with the Central Asian republics and major centers of Islamic Fundamentalism, such as Iran, through enhanced trade and investment links, expanding diplomatic ties, and Chinese assistance in critical development areas.[19]

China is highly suspicious of multilateral approaches to Asian security issues. Many strategists and political leaders believe that such approaches could ultimately (a) lead to the emergence of an anti-China coalition designed to restrict Beijing's defense modernization program; (b) come under the control of the United States and/or Japan and be used to increase the power of both states vis-à-vis China; and (c) facilitate, through inclusion in such dialogues, Taipei's efforts to achieve international recognition of a "one country, two governments" approach to the Taiwan issue, ostensibly as a prelude to formal Taiwanese independence. On a more practical level, many Chinese strategists also reportedly believe that multilateral approaches are largely irrelevant or potentially damaging to the resolution or management of key disputes in the Asia-Pacific, such as those between Russia and Japan. Because of such views, as well as China's long-standing opposition to "hegemonism," Beijing will likely continue to avoid a leadership role in multilateral fora, and generally prefer that such fora remain unstructured, informal, and merely consultative. It should be noted, however, that a minority of Chinese strategists look more favorably on multilateral security structures, primarily to constrain Japan in the future, or to handle specific kinds of subregional problems.[20]

The maintenance of reasonably good relations with the United States in particular serves many critical Chinese security goals: (a) the continued success of domestic economic reform through Western trade and investment, (b) the avoidance of excessive foreign pressures on

---

[19]These points are taken from Swaine (1995, p. 87).

[20]For further details, see Shirk (1994).

China's military modernization program, (c) the deterrence or avoidance of a more independent or assertive Japan, (d) the reduction of U.S. incentives to provide military assistance to Taiwan, and (e) assistance in the resolution of various critical issues of mutual concern, such as nuclear proliferation and the possible long-term reemergence of an expansionist Russia.[21]

Stability on China's Western borders is essential to national unity and the continued maintenance of a placid external environment conducive to economic reform. China's rapid economic development provides a strong incentive for the largely impoverished Islamic republics of the former USSR to expand trade and investment ties with Beijing, rather than to encourage the separatist activities of the Muslim minorities in China's Xinjiang province. Hence, Beijing's policy stresses efforts to strengthen the stakes of its Inner Asian neighbors in secular economic reform and close economic and political relations with China, not to project political influence across the border. This policy extends to Beijing's relations with Middle Eastern countries, especially Iran.[22]

In general, Beijing's diplomatic approach in these and other areas increasingly seeks to draw upon China's growing involvement in the dynamic Asian and world economy. Chinese strategy now stresses the use of economic appeals and/or leverage to build international support for diplomatic and security objectives and to make major powers aware that opposing core Chinese interests will likely undermine their own economic interests. This suggests a Chinese emphasis on the economic arena as an increasingly important domain of international competition.[23]

---

[21]These points are taken from Swaine (1995).

[22]For details on China's efforts to strengthen relations with all states of the "Middle East–Central Asian buffer zone," see Harris (1993, pp. 258–259). Some security analysts and U.S. officials believe Chinese overtures to countries such as Iran and Pakistan are also aimed at the creation of a strategic consensus designed to weaken U.S. influence in the Gulf and increase Beijing's diplomatic leverage vis-à-vis Washington.

[23] For an extended defense of this point, see Lampton (1994).

Local War, Active Peripheral Defense, and Power Projection

China's changed security, military, and economic environments have together generated five major requirements for Chinese defense policy, most focused on the Asia-Pacific region:

1. Increase China's overall global and regional stature, particularly through the acquisition of high-technology weaponry and the ability to "show the flag" beyond China's borders.

2. Deal with the uncertain future military postures of the United States, Japan, the ASEAN states, and perhaps India.

3. Maintain a credible threat of force toward an increasingly separatist-minded and economically potent Taiwan.

4. Improve Chinese military and diplomatic leverage over and access to nearby strategic territories claimed by Beijing, such as in the South China Sea, and to defend access to vital oceanic routes in the event of conflict.

5. Strengthen China's ability to deal with domestic social unrest and ethnically based border instabilities.[24]

These requirements have led to a significant transformation in China's strategic outlook, from that of a continental power requiring large land forces for defense against threats to its internal borders, to that of a combined continental and maritime power with a range of diverse domestic and external security needs. This shift is reflected in a broad-based defense doctrine comprising the central concepts of local war, active peripheral defense, and rapid power projection, mentioned above. Such notions are based, in turn, upon several new Chinese strategic principles and combat methods, such as an expanded definition of "strategic frontier" and the notions of "strategic deterrence" and "gaining the initiative by striking first."[25]

---

[24]These points are taken from Swaine (1995, p. 89).

[25]The revised Chinese principle of "strategic frontier" is intended to encompass the full range of competitive areas or boundaries implied by the notion of comprehensive national strength, including land, maritime, and outer space frontiers, as well as more abstract strategic realms related to China's economic and technological development. The principle of "strategic deterrence" was formulated to emphasize the nonviolent

These and other elements of China's post–Cold War defense doctrine, first enunciated by the Chinese leadership in the early and mid-80s, assume that limited or regional conflicts of relatively low intensity and short duration could break out virtually anywhere on China's periphery, demanding a rapid and decisive application of force. Such possibilities are suggested by points 3 through 5 above. Other elements of the doctrine, associated with points 1 and 2, assume that Chinese forces will eventually need to attain broader power projection and other capabilities sufficient to support China's long-term great-power ambitions.[26]

Taken together, such contingencies and aspirations demand the development of advanced weapons with medium- and long-range force projection, rapid-reaction, and offshore-maneuverability capabilities. This implies the creation of a smaller, highly trained and motivated, technologically advanced, and well-coordinated military force operating under a modern combined arms tactical operations doctrine utilizing sophisticated C^3I systems. In addition to conventional, well-equipped ground units, such a force should contain so-called rapid-reaction combat units with airborne drop and amphibious landing capabilities, as well as far more sophisticated air and naval arms, to perform both support and power-projection functions. To improve capabilities in the latter area, the Chinese now place a high priority on the development of air and naval electronic warfare systems, improved missile and aircraft guidance systems, precision-guided munitions, the construction of communications and early warning satellites, and in-flight refueling technol-

---

use of military power to deter war or to achieve political or diplomatic ends, in contrast to the traditional Chinese emphasis on the use of military forces in actual combat. An increased emphasis on gaining the initiative by striking first (rather than waiting for the enemy to strike) reflects the need to act quickly and decisively to preempt an attack, restore lost territories, protect economic resources, or resolve a conflict before it escalates. For further details on these and other less critical principles basic to China's post–Cold War defense doctrine, see Nan (forthcoming), pp. 7–14. Also see Godwin (1987, pp. 573–590).

[26]It should be noted that the specific structure and capabilities of China's military forces over the long term have yet to be determined by the Chinese leadership. Many issues remain under debate, including the optimal size and configuration of the navy, the number of major ground force units and their deployment, and the desired size and structure of the air force.

ogy.[27] Such a diverse set of military capabilities also requires a host of secondary features, including a more robust research and development capability, a more technologically advanced and quality-driven defense industry, and a highly professionalized, merit-based system of officer recruitment, promotion, and training.

Perhaps most important, these improved capabilities and support features also require relatively high levels of government spending on defense. As a result, total Chinese defense spending has increased by well over 10 percent per annum since at least 1990, and reports suggest that the Chinese leadership intends to maintain similar increases for at least the remainder of the 90s.[28] However, the exact level of Chinese military spending is a source of considerable debate, both within China and among foreign observers. Tables 8.1 and 8.2 provide a range of estimates for current Chinese military spending (based upon a variety of purchasing power parity (PPP)-based and non–PPP-based calculations) and projections of future spending levels.[29]

Such variations suggest the need for far more detailed information on defense-related revenues, as well as more precise methodologies for measuring military expenditures. At present, little more can be said about China's level of military spending beyond the obvious point that, *if trends since 1990 continue*, China will possess the

---

[27]In addition to such conventional improvements, China's new doctrine also calls for continued advances in the survivability, penetration, and retaliation capacity of its strategic nuclear force, which Beijing views as a critically important deterrent against the vastly superior conventional forces of the major industrialized states. This has focused on the qualitative improvement of intercontinental and intermediate-range ballistic missile capabilities and the creation of a more potent, albeit small, tactical nuclear arsenal for possible use in local war scenarios. In general, the emphasis has remained on quality over quantity. To improve its capabilities, China continues to conduct nuclear tests.

[28]The official Chinese defense budget increased over 15 percent in 1989 and 1990, nearly 14 percent in 1991, over 14 percent in 1992 and 1993, and 23 percent in 1994. Such increases reversed a decade of average negative growth rates.

[29]PPP estimates are based upon measurements of the approximate cost, in the United States, of a representative basket of Chinese goods and services. They employ an exchange rate calculated by comparing the prices in two countries of nontraded and traded goods, rather than a rate based upon the official exchange rate derived from estimates of traded goods and services only. The information presented in the following tables and paragraphs was taken from a forthcoming work by Swaine et al. on change in Asia and the sources of adversity for U.S. policy.

Table 8.1

Range of Estimates for Current Chinese Defense Spending

Source of Estimate	Year for Estimate	RMB Estimate (billions)	Dollar Estimate (billions)	Defense Share (percent)
Official exchange rate	1994	52.0	6.0	1.5
Official figures, International Monetary Fund, PPP based	1994	52.0	33.7	1.5
Official figures, University of Pennsylvania, World PPP	1994	52.0	75.9	1.5
International Institute for Strategic Studies	1994	100.0	45.0–55.0	3.3
Arms Control and Disarmament Agency	1990	—	55.0	—
Stockholm International Peace Research Institute '94	1993	258.7	45.0	8.6
Lowest combination (official/official)	1993–1994	52.0	6.0	1.5
Highest combination (SIPRI/Penn World)	1993–1994	258.7	377.6	8.6

financial capability to improve its force structure over the next 10 to 15 years to an extent that will cause growing anxiety among its neighbors. Absent the ability to define the size and disbursement of China's military budget more precisely, however, such a statement means little. To understand the long-term implications of China's defense buildup, one must therefore focus instead on improvements and changes occurring in areas other than military spending, such as force structure, key support infrastructures, C^3I capabilities, and Chinese threat perceptions and defense doctrines. For the purposes of this chapter, we shall focus on actual and planned naval and air acquisitions.[30]

NAVAL AND AIR ACQUISITIONS

Understandably, the focus of Beijing's arms and technology acquisitions in the 80s and 90s has been keyed to the creation of much

[30]Chinese threat perceptions and defense doctrines have already been discussed.

Table 8.2

Range of Estimates for Future Chinese Defense Spending

Source of Estimate	Dollar Estimate Circa 1993 (billions)	Implied Dollar Estimate for 2007 (billions)
Official exchange rate	6.0	20.0
Official figures, International Monetary Fund, PPP based	33.7	113.0
Official figures, Penn World PPP	75.9	254.0
International Institute for Strategic Studies	45.0–55.0	150.0–184.0
Arms Control and Disarmament Agency	55.0	184.0
Stockholm International Peace Research Institute '94	45.0	150.0
Lowest combination (official/official)	6.0	20.0
RAND estimate	38.0	125.0
Highest combination (SIPRI/Penn World)	377.6	1,262.0

SOURCE: Table 1, with 9-percent growth from 1993.

smaller but highly proficient naval and air forces and related capabilities required for rapid reaction and limited power projection.[31] The Chinese navy (known as the People's Liberation Army Navy [PLAN]) has added nearly 20 principal surface combatants (i.e., ships with at least 1,000 tonnes displacement) to its inventory since the mid-80s. It is acquiring a new class of destroyer (the Luhu, or Type 052), is upgrading versions of its mainstay Luda-class destroyers, and is developing a new class of missile frigate (the Jiangwei). These vessels possess significant missile capabilities (including Silkworm surface-to-surface antiship missiles, surface-to-air missiles, and anti-missile missiles), more sophisticated radar and fire-control capa-

[31]The following summary of actual and planned PLA force acquisitions is taken from Swaine (1994c). Sources used in that paper include IISS (1993, 1994, 1995), including the enclosed fold-out wall chart in the 1993 edition, entitled "Asia: The Rise in Defense Capability, 1983–1993"; Sharpe (1993); Wortzel (1993, 1994); "China Pursues Traditional Great-Power Status" (1994); Song (1989); Ball (1994); Grazebrook (1994); and Morgan (1994).

bilities, antisubmarine warfare (ASW) capabilities, and electronic countermeasures.[32] The PLAN is also developing new classes of resupply amphibious assault ships and missile patrol craft and is greatly expanding its number of mine warfare ships.[33]

China's total number of operational conventional submarines has probably dropped by over one-half, from 100 in the mid-80s to less than 50 today. The vast majority of the remaining submarines are outdated Romeo-class models from the late 50s. However, the PLAN is endeavoring to upgrade the quality of its submarine force, first by improving its Ming-class submarines,[34] and then eventually by domestically developing a new type of diesel-electric submarine to replace the Romeo and Ming classes. This indigenously built Song-class submarine includes advanced Russian, French, and possibly Israeli technologies. The first hull was launched in 1994. China has also purchased four sophisticated Kilo-class conventional submarines from Russia (of which two had been delivered by late 1995) and might eventually acquire up to 18 more.[35] Improvements are also under way in China's nuclear submarine fleet. These include plans to supplement (or probably partly replace) the PLAN's five Han-class nuclear attack submarines (SSN), and to produce an additional Xia-class nuclear ballistic missile submarine (SSBN), with an improved missile. There have also been persistent reports that the Chinese plan to construct or purchase one or two medium to large

---

[32]China's most modern, and largest, warship is the 4,200-tonne Luhu-class destroyer. One of an expected four Luhus had been built by early 1995. The improved Luda III destroyer is intended to replace 16 earlier, largely obsolete, versions. Two Luda IIIs were built by mid-1995. As many as six Jiangwei-class missile frigates could be in service by the end of 1995 (four had been commissioned by mid-1995), joining or partly replacing China's fleet of approximately 30 aging Jianghu-class frigates.

[33]The PLAN has added over 100 mine warfare ships of different classes to its inventory since the mid-80s, plus nine advanced Houxin-class missile craft.

[34]Overall, a tenth improved Ming-class submarine and one modified Romeo-class with an Exocet-type surface-to-surface missile have been commissioned thus far.

[35]There are also reports of negotiations with Russia to undertake the licensed production of Kilo submarines, which are well-suited for coastal waters and for such operations as naval blockades. See Kathy Chen, *Asian Wall Street Journal*, February 9, 1995.

(40,000- to 50,000-tonne) aircraft carriers, for deployment between 2010 and 2020 (IISS, 1993, p. 148; Caldwell, 1994; Eikenberry, 1995).

These acquisitions have brought about major improvements in several areas. For example, the operational range, firepower, and air-defense capabilities of principal surface combatants have been significantly increased, theoretically permitting many destroyers and frigates to operate with minimal air cover (Wortzel, 1994, p. 164). Moreover, as a result of these and other improvements, the PLAN has improved its capability to carry out more-sophisticated operations farther from shore and for longer periods. For example, the PLAN has conducted multiship task force operations and fleet exercises in recent years, involving surface, subsurface, and aviation assets. Operations included maintaining and breaking blockades, attacking pipelines, and locating and destroying enemy mines (Song, 1989, p. 226; Caldwell, 1994, p. 8).

In contrast to the above successes on the naval front, progress in improving China's air capabilities has been highly limited, although future plans remain ambitious. The PLA Air Force (PLAAF) continues to rely primarily upon obsolescent versions of the Soviet MiG-17, MiG-19, and MiG-21 (fighters and of the Soviet Tu-16 bomber. These are known, respectively, as the J-5, J-6, J-7, and H-6. (The vast majority—about 3,000—of China's fighters are J-6s.) The PLAAF also operates a small number (approximately 100) of the more advanced J-8 fighters, as well as about 500 low-performance Q-5 ground-attack aircraft. China is reportedly working to upgrade the H-6 into a multi-role interceptor naval strike aircraft capable of launching a cruise missile. But even if that effort is successful, production levels are expected to be extremely low (Wortzel, 1993, p. 21).[36] Production rates for the workhorses of the PLAAF fighter force are also very low (about 40 J-7s per year and only 12 J-8s), while production of the J-6 apparently ceased in the late 80s. Moreover, the J-8 (roughly comparable to a basic F-4 Phantom) continues to be plagued by engine and fuel-consumption problems and poor weapon systems. In general, Beijing's effort to develop an advanced indigenous fighter and combat aircraft industry has been largely unsuccessful, and there are few

---

[36]Wortzel states that the H-6 is only produced at a rate of four per year.

signs of a breakthrough occurring in at least the near future. (Wortzel, 1993, p. 17).[37]

To compensate at least partly for its airpower problems, China has purchased 26 Su-27 aircraft from Russia and may obtain another 48 by the end of the decade. Beijing is also reportedly arranging with Russia to manufacture under license an estimated 300 additional Su-27 aircraft.[38] The Su-27 is an all-weather, counterair fighter capable of operating from a carrier, with ramp-assisted takeoff. It has an integrated fire-control system; has look-down, shoot-down radar; and is refuelable in flight. The Chinese also intend to coproduce a hybrid version of the Israeli Lavi fighter (the F-10), possibly with both Israeli and Russian assistance.[39] This multirole fighter-bomber is almost identical to the F-16 in its interceptor and ground-attack roles. Successful incorporation of these two advanced fighters would significantly improve Beijing's air combat capabilities.[40]

Both the PLAAF and the PLAN are also making concerted efforts to develop or purchase various "force multipliers," especially airborne early warning and midair refueling capabilities. For example, foreign companies are reportedly outfitting the H-6 bomber with an air-to-air refueling system, probably to be used with the J-8 fighter. At least one expert has estimated that the PLA will likely attain an air refueling capability for about two squadrons (24 to 30) of aircraft within five years (Wortzel, 1994, p. 169). This would considerably expand the range of at least a small portion of China's fighter force beyond its current general operational distance of about 250 to 300 miles.

Several of the above improvements in advanced naval and air power and amphibious lift capabilities are related to China's ongoing effort to develop an array of rapid-reaction units (RRUs), as part of a doc-

---

[37]According to Wortzel, system integration and engine design and manufacturing are the key obstacles. For an excellent, highly detailed analysis of China's largely unsuccessful attempts to modernize its fighter force, see Allen et al. (1995).

[38]Initial reports on this deal (e.g., IISS, 1992, p. 148) incorrectly referred to the planned coproduction of the MiG-31, not the Su-27.

[39]Wortzel (1993, p. 17) states that over 1,000 Russian defense scientists have reportedly conducted defense industrial exchanges with China since 1991.

[40]China has also purchased ten IL-76 medium-range transport aircraft from Russia and 15 from Uzbekistan, mainly to improve the mobility and lift capabilities of China's RRUs.

trine of local war and peripheral defense. These RRUs include both several battalion-sized, infantry-based units assigned to various Chinese group armies and a special marine force attached to the PLAN, supported by amphibious assault ships and landing craft (Ball, 1994, p. 103).[41] At least one well-informed expert believes that Beijing today has the lift and transport capability to project about two division equivalents (about 25,000 troops) "a good distance" from its borders and can probably conduct an amphibious landing of *nearly* division strength (i.e., 10,000 to 14,000 men) "well away from its immediate territorial waters." (Wortzel, 1993, p. 23.)[42] Since 1993, amphibious landing exercises have reportedly involved a full regiment, rather than a single battalion, as was common before that year. This suggests that China might attain a full divisional amphibious landing capability within several years (Wortzel 1993, pp. 24–25). Most specialists on the Chinese military question such estimates, however, and insist that, although having made considerable progress in recent years, the PLA is far from attaining the capability to project two divisions or to land one division in an amphibious operation.

In addition to these improved infantry-based power projection capabilities, Beijing has also been improving its ability to conduct coordinated operations among several services, including those in support of amphibious landings. Since 1993, the PLA has conducted a series of increasingly sophisticated and extensive military exercises, many along China's eastern and southern coastlines. These exercises involve ground forces, surface and subsurface naval forces, marines, airborne drops in support of amphibious operations, and air forces.

China's major naval and air acquisitions are summarized in Table 8.3.

---

[41]The PLAN has recently acquired five Yukan-class amphibious transport ships or LSTs (each capable of carrying 200 troops) and the first of many Yuting-class tank landing ships.

[42]Wortzel estimates that China could probably employ both military amphibious landing ships and its sizable merchant marine to move the equivalent of a full Group Army (about 40,000 men) to conduct follow-up landing operations after the seizure of a beachhead.

Table 8.3

Recent and Planned Major Naval and Air Acquisitions

Branch	Type	Class	Recent	Planned[a]
Navy	Destroyer	Luhu	1	3
	Destroyer	Luda III	2	14
	Missile frigate	Jiangwei	4	2
	Submarine	Improved Ming	10	?
	Submarine	Modified Romeo	1	?
	Submarine	Song	0	?
	Submarine	Kilo	4	18[b]
	Mine warfare ships	Various	100	—
	Missile craft	Houxin	9	?
	SSBN	Xia	1	?
	SSN	Han	—	1–5?
	Aircraft carrier	40,000–50,000 tonne	—	1–2
	LST	Yukan	5	?
	Tank landing ship	Yuting	1	?
Air force	Fighter	J-7	40/year	—
	Fighter	J-8	12/year	—
	Fighter	Su-27	26	348[c]
	Fighter	F-10	—	?[b]
	Bomber	H-6	4/year	
	Transport	IL-76	25	?

[a]These numbers indicate quantities of weapon systems under consideration or already determined. Question marks indicate uncertainty about a planned quantity.

[b]An unknown quantity to be coproduced.

[c]Includes about 300 to be coproduced.

IMPLICATIONS FOR THE UNITED STATES

China's security policy objectives and program of defense modernization could eventually pose a major challenge to U.S. interests and capabilities in Asia. On the broadest level, the attainment of a significant Chinese blue-water naval capability, along with limited but significant amphibious landing and airborne drop capabilities, could severely undermine confidence in the ability of U.S. forces to continue to serve as the ultimate guarantor of peace and stability in the region. For example, the ability of Chinese naval forces to transit, on a frequent basis, and in significant numbers, critical maritime areas and strategic lines of communication (SLOCs), such as the South

China Sea and the Strait of Malacca, would almost certainly alter regional perceptions and raise anxieties. As a result, many Asian nations might be forced to accelerate their existing programs of military modernization or otherwise alter key aspects of their current security policies in ways that do not serve U.S. interests. Such adverse changes might eventually prompt major alterations in the size and/or deployment of U.S. air and naval forces in the region.

On a narrower level, steady improvements in China's economic and military capabilities could lead to more assertive Chinese behavior toward specific flashpoints, such as Taiwan, Hong Kong, the Spratly Islands, the Korean peninsula, and even the Indian Ocean. Such behavior would likely produce major diplomatic tensions and perhaps even armed confrontations that could lead to the deployment of U.S. air and naval forces. Such assertive Chinese behavior could also result from domestic political factors, most likely involving attempts by the leadership in Beijing to play the "nationalism" card. For example, Chinese leaders might attempt to provoke a confrontation with Taiwan, or over the Spratly Islands, to unify a fragmented and conflictual successor leadership, distract popular attention from internal woes, such as an economic crisis, or strengthen military support for (or control over) a weak, insecure regime. Such a confrontation could also be used by individual groups or factions within the civilian or military leadership to further their narrow political or institutional interests.[43]

Other, less conflict-oriented Chinese outcomes demanding the use of U.S. forces could also occur, of course. For example, as suggested in the introduction, domestic economic and political chaos and social unrest in China could necessitate the deployment of U.S. forces to control refugee flows, to evacuate American citizens or friends from unstable or dangerous areas, or to provide relief to starving or injured citizens. Such extreme outcomes also present the possibility of a loss of control over nuclear weapons by the Chinese government, perhaps leading to highly unstable and dangerous situations requiring U.S. military intervention.

---

[43]For example, the PLAN might seek a more assertive stance toward Taiwan to justify stronger central support and bigger budgets for the development of a modern, technically proficient, combat-ready blue-water capability, as it has done with policy toward the Spratlys.

One should not assume, however, that one or more of these nightmare scenarios will inevitably follow from current Chinese trends.
First, the Chinese leadership is fully aware that highly provocative
military behavior toward the region (e.g., regarding Taiwan and the
South China Sea) would greatly threaten, if not entirely overturn, the
reform-based domestic and foreign policy strategy that has been the
key to China's diplomatic and economic successes for nearly 20 years
and would, in general, destabilize the entire Asia-Pacific region.
Specifically, such Chinese behavior would likely (a) destroy the Sino-
American détente that has underlain the stability and security of East
Asia since the 70s, increase the U.S. military presence in Asia, and
perhaps even precipitate armed conflict between the United States
and China; (b) cause virtually every Asian country to doubt China's
avowed commitment to the peaceful resolution of its territorial and
other disputes with its neighbors, thus perhaps resulting in a major
acceleration in regional military acquisitions and other instabilities;
(c) disrupt trade and investment flows throughout the region; (d)
undermine growth and reform in China's coastal provinces and perhaps in China as a whole; and (e) probably produce political unrest
within China.

Moreover, continued rapid rates of Chinese growth will almost certainly lead to even closer economic, cultural, and diplomatic ties
with the Asia-Pacific region, thus strengthening elite interests within
and outside China favoring even greater cooperation with neighboring states. Indeed, Beijing may rely increasingly, perhaps primarily,
on its growing economic links to the region, rather than on its future
military prowess, to attain its diplomatic and strategic ends.[44] All this
suggests a continued cautious Chinese approach toward the United
States, combining elements of both cooperation and competition.

Second, many factors militate against a complete breakdown of
internal order and/or the fragmentation or breakup of China. The
PLA will likely continue to serve as a guarantor of national unity and
central government control after the death of Deng Xiaoping, barring

---

[44]Whether China relies more heavily in the future on its military instruments or its
economic strengths to attain many of the above diplomatic and strategic goals may
depend in large part on how much influence the senior military leadership is able to
exert over foreign policy in a post–Deng Xiaoping setting. For an expanded discussion
of this issue, see Swaine (1995).

a near complete disintegration of the central leadership. Internally, the Chinese military is less regionally divided and both less capable and less inclined to intervene autonomously in elite politics than at any time in its recent past. It also does not exhibit the internal ethnic divisions that plagued the Soviet military and contributed to that force's eventual fragmentation.[45] In addition, China contains a highly homogeneous civilian ethnic population and a firm sense of national identity. In contrast to the situation in the former Soviet Union, well over 90 percent of China's population belongs to a single ethnic group, the Han Chinese, most of whom view themselves as members of a single nation-state.[46] Moreover, despite the emergence of regional centers of economic power under the reforms, the Beijing government retains important controls over the Chinese economy and over local political leaders. Perhaps even more important, many key provincial economies are increasingly dependent upon other provinces for continued growth. In short, few incentives exist for provinces or regions of China to break away from the center, or from one another.[47]

The above should not, however, lead one to preclude the possibility of nationwide chaos in the future. A prolonged and severe economic downturn, perhaps combined with a protracted succession struggle after Deng Xiaoping's death, could produce a vicious circle of leadership conflict, policy paralysis, and social unrest that might split the military and threaten internal order throughout China.[48] The implications of such a scenario for the control of China's nuclear arsenal are difficult to assess. Very little is known about China's system of controls over its nuclear forces, including the nature of safeguards against accidental or unauthorized use. However, certain basic features of the PLA command and control system and the deployment and size of China's nuclear force suggest that the prob-

---

[45]For a detailed discussion of the propensity for Chinese regional forces to become involved in future political struggles in China, and a comparison with the Soviet case, see Swaine (1994a, pp. 59–84).

[46]Although the vast majority of China's minority ethnic groups are concentrated primarily in sensitive border areas, they are usually heavily interspersed with Han Chinese and under the tight control of locally garrisoned PLA forces composed almost entirely of Han Chinese.

[47]For further details, see Swaine (1995).

[48]This and other scenarios are described in considerable detail in Ibid.

lem of "loose nukes" under chaotic conditions would be far less serious than in the case of the former Soviet Union.[49]

Third, dire projections concerning future Chinese defense capabilities may be unwarranted because the Chinese leadership may prove unable to maintain existing levels of defense spending over a long period. Beijing faces a major, and growing, fiscal crisis, as public revenues decline and expenditures mount.[50] A radical restructuring of the Chinese tax system will be necessary to remedy this problem and ensure high levels of military funding. Although the central government is currently moving in such a direction, the obstacles remain enormous, and the effort could ultimately fail, thereby likely ensuring a steady decline in central government capabilities.[51]

Finally, even under the most stable and supportive political, economic, and social circumstances, China will need to overcome many serious technical and organizational problems plaguing its military modernization program before it can attain the kind of regional capabilities or carry out many of the types of assertive regional behavior outlined above. For example, by most accounts, the PLA continues to suffer from very poor command and control capabilities for its air, naval, and land forces (including rapid-reaction units), despite recent improvements. As many analysts have pointed out,

---

[49]First, China possesses far fewer nuclear weapons than did the former Soviet Union, and virtually all of these weapons are located outside ethnic minority areas. These include an arsenal of approximately 300 deployed nuclear warheads and about 150 additional tactical nuclear weapons that are available but not deployed. The former are divided into a triad of land-based missiles, bombers, and a few submarine-launched missiles. It is estimated that only four land-based missiles are ICBMs capable of striking the continental United States. These are reportedly located in Henan Province, far from ethnic minority regions. Other land-based missiles of considerably shorter range are deployed in various provinces throughout China, many (if not most) outside ethnic border regions. Second, China's nuclear arsenal is under the direct control of the top leadership in Beijing, through the communist party's Central Military Commission. The units of the Second Artillery Corps, responsible for the maintenance and deployment of all land-based nuclear missiles, are directly under this party organ. Third, tactical nuclear weapons are probably stored at a central storage site or at a very small number of regional sites. They are not deployed among troops in the field. For further details see Norris, Burrows, and Fieldhouse (1994).

[50]For example, see Wong (1991).

[51]One alternative is to expand China's nontax sources of revenue by increasing arms sales. This will probably not generate sufficient funds, however, and could precipitate both increased tensions with the West and an internal leadership dispute.

such a fundamental problem can *greatly* diminish the apparent advantage gained by both quantitative and qualitative advances in force levels or weapon systems.[52]

In addition, the PLAN continues to exhibit several major deficiencies, despite significant recent advances. Many of its destroyers and frigates still lack antisubmarine capabilities and modern fire-control systems. Moreover, all of these vessels lack a long-range surface-to-air (SAM) system and effective defense against antiship missiles. In addition, virtually all of China's submarines are of obsolete design and thus relatively easily detectable by modern sonar equipment. China's five nuclear-powered attack submarines, launched in 1972 and 1977, are probably not operational, and some may even have been scrapped (Morgan, 1994, p. 33). Moreover, it will take many years for China to obtain (or manufacture) and effectively operationalize a significant number of improved or new nonnuclear submarines, such as the Ming or Song classes or the Russian Kilo class noted above. Finally, China's many fast-attack naval craft and minesweepers are defensive, with limited operational range (Morgan, 1994, p. 34).[53]

The PLAAF suffers from even more difficulties than the PLAN, including a very weak air-to-ground attack capability, a generally poor missile inventory, insufficient combat training and logistical support, poor C^3I capabilities, a largely defensive fighter force structure, and overall obsolescence in airframe design and key technologies. Moreover, the fighter-bomber fleet has a combat radius of only about 280 nautical miles, and the PLAAF is probably still several years from developing a fully operational and extensive aerial refueling capability. Also, PLAN pilots are probably the only airmen trained to navigate over water at extended range (Wortzel, 1993, p. 17; Godwin, 1993, p. 20).[54]

We should also mention that studies conducted by RAND researchers of PLAAF fighter production and retirement and attrition

---

[52]For example, Godwin (1993, p. 19), and Eikenberry (1995, p. 87).

[53]Also see a Hong Kong publication on the PLAN entitled "Chinese Communist Naval Forces" (1994, pp. 30–31).

[54]Also see Allen et al. (1995), especially pp. 163–165. This study systematically examines the many obstacles the PLAAF faces in its effort to modernize.

rates suggest that, if the Chinese attempt to keep their fighter force at current levels, over 1,500 additional aircraft would probably be required between 1995 and 2005. Assuming that J-8 production could be more than doubled, to 40 units per year, J-7 production would still need to exceed 140 per year, a major increase. This would result in a force composed of almost 60-percent J-7 and J-8 aircraft, capable of intercepting aircraft over China, but still inadequate for power projection and ground attack, especially without an extensive aerial refueling capability.[55]

The acquisition by China of only another 26 or 48 Su-27s or comparable aircraft during the remainder of the decade will certainly not resolve these basic problems.[56] Moreover, it will take many years for China to attain significant production levels for the Su-27, and even at peak production, it is estimated that China will likely manufacture only about 50 aircraft per year through the above-mentioned coproduction arrangement with Russia (Tai, 1993). Equally important, it will take pilots, ground crews, and logistics personnel several years to master the required technologies and operational features of the Su-27.

The above factors suggest that China will likely require a significantly long time (i.e., *at least* 15 to 25 years) to attain a truly modern force structure and operational capability capable of challenging the U.S. military presence in the region. Yet, even modest improvements in China's power-projection capabilities could generate serious instabilities in specific subregions of Asia, such as the South China Sea and Taiwan.[57] Much will depend on future Chinese economic growth rates and defense spending levels, the outcome of efforts to overcome the above technical military problems, the composition and outlook of the post-Deng Chinese leadership, and, of course, the behavior of external countries and territories. In short, Chinese

---

[55]Interview, RAND researcher.

[56]According to Klintworth (1994, p. 14), USAF intelligence specialists believe that the PLAAF's 26 Su-27s have little operational significance and that even 74 aircraft would have a relatively modest impact on the regional balance of power.

[57]Indeed, the marked increase in tensions between Taiwan and China that resulted from the June 1995 visit of Taiwanese President Lee Teng-hui to the United States have significantly increased the chances of Beijing deploying more advanced Chinese naval and air assets against the island over the near to medium term.

defense capabilities and behavior will almost certainly continue to be affected by a series of highly dynamic changes. Understanding the nature of these changes, and their impact upon the security environment in Asia and beyond, will constitute a major challenge for the future.

REFERENCES

Allen, Kenneth, Glenn Krumel, and Jonathan D. Pollack, *China's Air Force Enters the 21st Century*, Santa Monica, Calif.: RAND, MR-580-AF, 1995.

Ball, Desmond, "Arms and Affluence: Military Acquisitions in the Asia-Pacific Region," *International Security*, Vol. 18, No. 3, Winter 1993/1994, pp. 78–111.

Caldwell, John, *China's Conventional Military Capabilities, 1995–2004: An Assessment*, The Center for Strategic and International Studies, Washington, D.C., 1994.

"China Pursues Traditional Great-Power Status," *Orbis*, Spring 1994, pp. 157–175.

"Chinese Communist Naval Forces," Hong Kong, June 1993, translated in FBIS-CHI-002, January 5, 1994, pp. 30–31.

Eikenberry, Karl W., "Does China Threaten Asia-Pacific Regional Stability?" *Parameters*, Vol. XXV, No.1, Spring 1995, pp. 83–103.

Garrett, Banning, and Bonnie Glaser, "Multilateral Security in the Asia-Pacific Region and its Impact on Chinese Interests: Views from Beijing," *Contemporary Southeast Asia*, Vol. 16, No. 1, June 1994, p. 14.

Godwin, Paul H.B., "Changing Concepts of Doctrine, Strategy, and Operations in the People's Liberation Army 1978–87," *The China Quarterly*, No. 112, December 1987, pp. 573–590.

_____, "The Use of Military Force Against Taiwan: Potential PRC Scenarios," in Parris H. Chang and Martin L. Lasater, eds., *If China Crosses the Taiwan Strait: The International Response*, Lanham, Maryland: University Press of America, 1993, p. 20.

Grazebrook, A. W., "The Year at Sea: Regional Naval Growth Continues," in *Asia-Pacific Defense Reporter 1994 Annual Reference Edition*, Vol. XX, No. 6/7, December 1993/January 1994, pp. 6/7.

Harris, Lillian Craig, *China Considers the Middle East*, New York: I. B. Tauris and Co., Ltd., 1993.

IISS—See The International Institute for Strategic Studies.

The International Institute for Strategic Studies, *The Military Balance, 1992-1993*, London, 1992.

_____, *The Military Balance, 1993–1994*, London, 1993.

_____, *The Military Balance, 1994–1995*, London, 1994.

_____, *The Military Balance, 1995–1996*, London, 1995.

Klintworth, Gary, "China: Myths and Realities," *Asia-Pacific Defence Reporter*, April–May 1994.

Lampton, David M., "China and the Strategic Quadrangle: Foreign Policy Continuity in an Age of Discontinuity," in Michael Mandelbaum, ed., *The Strategic Quadrangle: Japan, China, Russia, and the United States in East Asia*, New York: Council on Foreign Relations Press, 1994.

Morgan, Joseph R., *Porpoises Among the Whales: Small Navies in Asia and the Pacific*, East-West Center Special Report No. 2, March 1994.

Nan Li, "War Doctrine, Strategic Principles and Operational Concepts of the People's Liberation Army: New Developments (1985–1993)," *The China Quarterly*, London (forthcoming).

Norris, Robert S., Andrew S. Burrows, and Richard W. Fieldhouse, *Nuclear Weapons Databook: Vol. V, British, French, and Chinese Nuclear Weapons*, Natural Resources Defense Council, Inc., Boulder, Colo.: Westview Press, 1994.

Shambaugh, David, "China's Security Strategy in the Post-Cold War Era," *Survival*, Vol. 34, No. 2, Summer 1992, pp. 88–106.

_____, "Growing Strong: China's Challenge to Asian Security," paper prepared for presentation to the Fourth Annual Staunton Hill Conference on the People's Liberation Army, August 27–29, 1993.

_____, "The Insecurity of Security: The PLA's Doctrine of Threat Perception Toward 2000," *Journal of Northeast Asian Studies*, Vol. XIII, No. 1, Spring 1994, pp. 3–15.

Sharpe, Richard, ed., *Jane's Fighting Ships 1993–1994*, Coulsdon, England: Jane's Information Group Ltd., 1993.

Shirk, Susan, "Chinese Views on Asia-Pacific Regional Security Cooperation," *Analysis*, National Bureau of Asian Research, Vol. 5, December 1994, pp. 5–15.

Song, Yann-Huei (Billy), "China and the Military Use of the Ocean," *Ocean Development and International Law*, Vol. 20, 1989, pp. 213–235.

Swaine, Michael D., "Chinese Regional Forces As Political Actors," in Richard H. Yang, et al., eds., *Chinese Regionalism: The Security Dimension*, Boulder, Colo.: Westview Press, 1994a, pp. 59–84.

_____, "The Modernization of the People's Liberation Army: Prospects and Implications for Northeast Asia," *Analysis: National Bureau of Asian Research*, Vol. 5, No. 3, October, 1994b.

_____, "Arms Races and Threats Across the Taiwan Strait," a paper prepared for a conference on Chinese Economic Reform and Defense Policy, cosponsored by the International Institute of Strategic Studies and the Chinese Council of Advanced Policy Studies, Hong Kong, July 8–10, 1994c.

_____, *China: Domestic Change and Foreign Policy*, Santa Monica, Calif.: RAND, MR-604-OSD, 1995.

Tai Ming Cheung, *Far Eastern Economic Review*, July 8, 1993.

Wang Jisi, "Comparing Chinese and American Conceptions of Security," North Pacific Cooperative Security Dialogue: Research Programme, Working Paper Number 17, York University, September 1992.

Wong, Christine, "Central-Local Relations in an Era of Fiscal Decline: The Paradox of Fiscal Decentralization in Post-Mao China," *China Quarterly*, No. 128, December 1991, pp. 691–715.

Wortzel, Larry, "Engaging the New World Order: Where is the PLA?" paper prepared for the Fourth Annual Staunton Hill Conference on the PLA, August 27–29, 1993.

_____, "China Pursues Traditional Great-Power Status," *Orbis*, Spring 1994, pp. 157–175.

Wu, Yu-Shan, "Taiwan in 1993: Attempting a Diplomatic Breakthrough," *Asian Survey*, Vol. XXXIV, No. 1, January 1994, pp. 46–54.

Chapter Nine

MIDDLE EAST

Graham Fuller

INTRODUCTION AND BACKGROUND

Recent developments in the Middle East have demonstrated a considerable lessening of potential interstate conflict in the Middle East and an increase in internal state turmoil in a number of significant countries. The most positive event was the September 13, 1993, signing of an accord between the Palestine Liberation Organization and Israel in Washington and the 1994 peace treaty between Israel and Jordan. The most negative events continue to be the continuing inability of Western policies to bring down the regime of Saddam Hussein in Iraq and the instability related to those efforts; increased tensions in U.S.-Iran relations because of Iranian opposition to the peace process and its alleged nuclear weapon program; the severe deterioration of the internal situation in Algeria, where an Islamic fundamentalist takeover becomes an ever more likely possibility; the continuing deterioration of the security situation in Egypt; and the continuing instability in Yemen that has the potential of turning into a broader geopolitical struggle.

This chapter reviews the major geopolitical trends of the Middle East and analyzes their likely course of development and their implications for U.S. policy interests and Air Force planning, where relevant.

THE PEACE PROCESS

The signing on the White House lawn of an accord between the PLO and Israel represents the greatest breakthrough in the long history of the Palestinian-Israeli conflict. For the first time, both parties, Israel and the PLO, now know what they must do to achieve peace and are directly talking on all issues. Realism marks the agreement. The strategic outlook for peace is now extremely positive: It is highly likely that a permanent settlement can be reached between the two sides. This judgment in based on the clear indications that both parties are no longer evading the realities—as had so long characterized previous efforts to reach agreement. Tactically, the process is still very messy and complex; bargaining is fierce, and both the Israeli and Palestinian populations contain elements opposed to the process, some of which are even are willing to use violence to stop it. Hence, the process of getting there will remain a tricky one, but it is unlikely to be significantly derailed over the long term. Today, there is no longer any pretense among Arab states that any group of states is either interested or willing to engage in military conflict with Israel.

The Israeli-Palestinian agreement was followed by the signing of a peace treaty between Jordan and Israel. The peace between Israel and Jordan cannot realistically be limited to a "cold peace." Cold peace between Egypt and Israel was basically the result of the severe isolation of Egypt in the Arab world after Sadat's unilateral signing of a peace treaty. Today, signatories of peace treaties with Israel are not isolated. The Palestinians, in particular, are intimately involved with Israel. They require jobs in Israel; their economies need each other. A cold peace would be disastrous to a future Palestinian state. Even if negotiations between the Palestinians and Israel founder over certain issues, such as the settlers or Jerusalem, there is no road back. Serious tensions could then emerge between the two parties, but they are unlikely to reverse the process. Both sides need a working peace.

Palestinian and Jordanian willingness to reach independent agreement directly with Israel has further reduced the power of Syria to manipulate the process to its own interests. The possibility of military conflict in the Arab-Israeli arena is thus lower than ever before. For the first time in half a century, a comprehensive settlement may now be on the horizon. Egypt has long since made peace; Jordan and

Israel have signed a peace treaty; and the Palestinians have taken control of Gaza and Jericho. Syria is thus isolated in the Arab world with no potential partner available to strengthen a negative position. Syrian President Hafiz al-Assad is moving slowly and grudgingly, but he has almost surely made his basic strategic decision to reach a peaceful settlement—even if the posturing and haggling will go on for a long time.

Syria at this stage is not only negotiating over the Golan but also over the future power constellation in the Levant that will deeply affect Syrian interests. Syria has basically been the loser in the process; it has lost Soviet arms and aid; it can no longer manipulate the PLO and has only marginal hold over Palestinian radicals; a settlement will lessen its claim to a grip over Lebanon; its ideological alliance with Iran will be diluted; and it will no longer be the leader of the rejectionist camp as it has been for so many years. The major pause for thought is whether Syria finds all this too high a price to pay. But its other options are limited: It can only find potential support for rejectionism either in Iran or Iraq; the ability of either of these states to help form a meaningful rejectionist front is quite limited. One would have to posit a new axis of Syria, Iran, and Iraq, which does not seem long viable.

A peaceful settlement of the Syrian-Israeli borders could well involve United Nations (UN) or other international forces to guarantee it. This could be a process in which the U.S. military would become involved—most importantly because of U.S. security commitments to Israel, in which policing of the agreement would be seen as politically important to both Israel and the United States. A U.S. or UN military presence would be largely symbolic. The policing of the agreement would be very unlikely to involve potential hostilities: Israeli-Syrian agreements in the past have been strictly honored. In addition, the agreement is the critical linchpin to the long-range solution of the Arab-Israeli problem and hence in the U.S. and Israeli interest. Conservative forces within Israel that oppose any concessions on the Golan, however, strongly oppose any U.S. or UN role there that might facilitate the turnover of the territories back to Syria.

One of the most important features to watch in the unfolding of the peace process is the reaction of Islamic fundamentalist organizations. Nearly all fundamentalist groups have declared their opposition to the peace process, their refusal to recognize the existence of Israel, and their long-term rhetorical goal of ultimately liberating all of Palestine from Israeli control. The fundamentalist Palestinian Hamas organization, for example, is still conducting operations of political violence against Israeli military officials and sometimes even civilians. As the peace process becomes ever more a reality, and the armed struggle of the intifada days fade, however, Hamas is being forced to conform somewhat to the new realities. It now seeks to participate in Palestinian elections. It is required to cut back on its paramilitary operations since those operations for the first time begin to have an effect on the Palestinian entity and a population with a new stake in the new arrangements. Hamas seems more interested in a long term political role as an opposition party than it is in mere perpetuation of terrorist acts.

Other fundamentalist organizations in the Arab world, including the Shi'ite Hizbollah in Lebanon that is linked to Iran, also have to face the new realities. Even the regime in Tehran itself faces the prospect of its only close ally, Syria, coming to terms with Israel. Fundamentalist groups may therefore retain some of their old rhetoric, but in reality will be forced to adopt more pragmatic positions if they wish to contend in serious politics. Only fringe terrorist organizations will likely be the sole remaining groups maintaining the old terrorist agenda. This development is a very important strategic evolution in the Middle East—still only nascent, but deserving of close attention. Terrorism is hardly dead in the Middle East, but it will be less spawned by the Arab-Israeli issue and will relate more closely to internal political struggles in the future.

EGYPT: THE CONTINUING STRUGGLE AGAINST FUNDAMENTALISM

The internal security situation in Egypt has continued to deteriorate over the past year. The number of attacks by militant fundamentalist (Islamist) groups has risen, while the government's crackdowns and arrests have also been continuously on the rise. While the government is not yet in danger of losing control of the situation, public dissatisfaction with the harsh measures is rising. Most citizens in Egypt have little sympathy for the acts of violence and terrorism perpetrated by these Islamist groups—most notably al-Gama'at al-

Islamiyya (Islamic Associations)—but at the same time believe that their actions have some basis of justification in the face of widespread corruption, lack of political freedoms, the declining capabilities of the government to meet pressing social and economic needs and to stem police violence and torture, and the increasing isolation of the government.

Worse, the Islamists have gradually gained a monopoly over nearly all expression of opposition. There are no other meaningful opposition groups in Egypt; most of them have been repressed in the past, otherwise marginalized, and generally lack mass following. The Islamists today represent the only real mass party, even if they are not organized as such. Although the major issues of grievance do not involve Islamic or religious issues per se, the Islamists are the only ones at this point to press successfully demands for improvement in living conditions, jobs, and an end to corruption.

The policies of the Mubarak government seem almost guaranteed to produce yet greater instability in the future. The government is continuing its policies of severe repression, drawing criticism from international human rights organizations. At the same time, there has been almost no effort to open up the political system in which a variety of opposition groups could flourish and offer competition to the more radical Islamists. The government has grown more isolated; access to the president is narrowing; and the regime is undertaking no creative actions to change the situation on the ground. The regime has also chosen to deliberately polarize the situation, forcing most of the intellectuals and the elite to choose between the Islamists and the regime; presented with this choice, most will choose to back the regime in the confrontation. The Islamists thus may not win over a majority to support them but may find increased segments of the population not offering active support to the government either. Under these circumstances, the government could find its support "hollowing" as the Islamists gain strength and come to represent one of the few sources of "legitimacy" in the country.

Unless the government sharply improves its tactics and opens up the system to greater external participation and to constructive criticism, the Islamists are likely to be the major beneficiaries of the deteriorating situation. Raising the standard of living and improving the overall economic situation is important, but the problem is no longer one of mere economics: It involves regime legitimacy as well, a commodity that is increasingly threatened in Egypt. A victory of the Islamic Liberation Front (FIS) in Algeria would greatly increase the political pressure on the Mubarak regime in Egypt.

The importance of Egypt to the United States cannot be exaggerated. It is the key U.S. ally in the Arab world. It is the cornerstone of the peace arrangement with Israel and has played the key role in the region in forwarding the peace process among the other Arab states over the past 15 years. It is also the single most important Arab state whose policies have immense impact on all other states in the region. Were Egypt to be taken over by an Islamist government in the next few years, the impact on the region would be very serious. It could place in jeopardy—but not automatically destroy—the Camp David peace treaty with Israel. It would have a negative impact on the peace process: probably the Palestinians and Jordanians would forge ahead, but Syria might well reconsider. Opposition to land for peace in Israel would sharply increase. Egypt would be in a position to assist other Islamist movements in the region. The psychological impact of Egypt "going Islamist" would be even more important. A reordering of regional relationships would be under way. An Islamist Egypt would also likely terminate the U.S. MFO force in the Sinai.

Yet, apart from more economic aid, and political advice on the internal situation—unwelcome to a regime that considers itself to be the better judge of the internal dynamics of the situation—there is little that the United States can do. If there are increasing internal disorders, it is difficult to imagine a military role for the United States in the middle of a civil war that involves neither regions nor separate ethnic communities. A peacekeeping or peacemaking role is therefore unlikely, although the United States would be open to maximum cooperation in any international plan designed to stabilize Egypt. Humanitarian operations in the event of a collapse of government or general urban chaos are a greater likelihood. Nation-building, however, would not be a necessary task where there is a long tradition of government and bureaucracy that would survive to some degree, despite political chaos.

Egypt faces no serious external military threat, except potentially Israel. In the event of a serious breakdown of relations between Israel and Egypt under some kind of Islamist regime in Cairo that led to hostilities between the two countries, it is quite conceivable that the United States could become involved in peacekeeping between enemy lines. Such a threat is not now on the horizon. The situation could change rapidly with the emergence of an Islamist regime, however, whose policies and style cannot readily be foreseen now.

DETERIORATION OF ALGERIAN STABILITY

The political situation in Algeria has continued to deteriorate as the struggle between the regime and the Islamic fundamentalist movement FIS sharpens. Over 4,000 people have now died in the conflict in the past two years as the civil war has spread. The military is no longer able to control the conflict, while Algerian society becomes increasingly divided on the issue. The government has changed hands several times in the last two years in a desperate search for a political and military strategy that can put the situation to rest.

The FIS itself has polarized, with more radical and violent elements within it coming to the fore as the conflict sharpens. A new President, Liamin Zeroual, is looking for ways to open dialog with the fundamentalists, but faces considerable opposition himself from within the ranks of the military, the ranks of the French-speaking elite, and the Berber minority, among others. Even if the government should be able to open a dialog with the FIS, there is a possibility that other militant hard-line Islamist organizations would not accept it if they believe that they can win the armed struggle without negotiation.

Important strategic issues are at stake here:

- The armed struggle is debilitating the country and dividing it politically. Whatever the outcome, Algerian politics are likely to remain turbulent for a long time.

- If the FIS should come to power, large numbers of Algerians opposed to an Islamist regime will seek to flee the country; this process has already begun as large numbers of professionals who see themselves as potential targets of Islamist violence have left. Large waves of refugees will present security and other problems to neighboring states, such as Morocco and Tunisia. Southern Europe, especially France, will be the main goal of Algerian refugees, which already has deeply worried the French government. The French government views the potential collapse of the Algerian government in the face of the FIS as a profound blow to French prestige, influence, and interests. It is now uncompromising in its support for the Algerian government and opposes any inclusion of the FIS in power.

- Algeria is increasingly the source of natural gas for Western Europe, especially Spain. By 1997, Spain will receive the bulk of its gas from Algeria; political turmoil or the "political use" of gas by an Algerian regime could considerably affect the European economy—although there is no indication so far that the FIS would have any such intentions of doing so. A victorious FIS would desperately need such proceeds for its own social programs.

- The FIS has now been much more radicalized as a result of being denied its legitimate election victory in December 1991 by the January 1992 military takeover. Vicious fighting and harsh security measures have further marginalized the FIS moderates, making a possible future FIS government much more unsettling in the region than would previously have been the case. Such a regime, if it comes to power, by the symbolism of its victory alone would greatly strengthen other fundamentalist movements in Morocco, Tunisia, Libya, and Egypt.

- Whatever the outcome, the wounds will not be healed for a long time. Algeria is likely to bear the scars and to adopt a generally prickly nationalist role in the region, including long-nurtured ambitions (by earlier regimes) to become a nuclear power and one of the regional great powers of Africa. So far such ambitions are almost strictly theoretical, since there are no indications the country has any such capabilities, and its funds are extremely limited. Over the longer term, however, this question must be closely watched. Algeria is the power to be reckoned with in the region over the longer term.

- A hostile Algeria could readily develop a missile capability against Western Europe. The distance from Algiers to the country's southernmost city is longer than the distance between Algiers and Paris.

The Algerian situation represents a situation in which a humanitarian, or peacekeeping role involving the United States is unlikely, but could not be completely ruled out. Any external intervention in the country to stop the civil war, restore order in the capital and major cities, protect oil and gas installations, and restore urban services in the highly populated capital city would first and foremost involve the French, perhaps in a NATO capacity. But France is a highly volatile symbol of past colonialism, and the FIS would not regard it as a neutral force or arbiter. The United States might not be regarded as sharply different from the French in the eyes of the FIS. Since the security of the Mediterranean region and NATO allies is involved, however, NATO involvement remains a possibility in a peacekeeping capacity. External intervention would not be possible in the case of civil war.

IRAQ: SADDAM'S DECLINE AND THE KURDISH DILEMMA

Despite heavy international sanctions, Saddam Hussein has still not fallen. The Iraqi economy continues its downward plunge; the Iraqi dinar has been devalued to at least a thousandth of its prewar value; and the regime's authority is dwindling outside Baghdad. Saddam's own behavior grows more desperate; there have been coup attempts within the army; and even members of his own clan have been arrested and executed. But he has not fallen yet. As long as Saddam remains in power, especially under deteriorating conditions, the chances of his making another foolish move—out of bravado and desperation—increase. This is what he did in October 1994 when he moved his forces toward Kuwait. It is entirely possible that Saddam will strike out again to try to save his domestic situation; an attack on Iraqi Kurdistan to the north would be one likely scenario, and a symbolic attack against Kuwait would not be entirely out of character for this tyrant who possesses exceptionally poor judgment.

Saddam would most likely move against the no-fly zone of the north, south of the area of Operation Provide Comfort currently manned by United States and other international personnel:

• Saddam has military forces near the region and could move relatively quickly and regain control of large areas of Kurdistan.

- His hope would be that it would be difficult for the allies to dislodge him short of undertaking a major war.

- He would be perceived by most countries as simply trying to reestablish control over his own sovereign territory—which would meet with the understanding of most countries in the region.

- He would claim he is attempting to preserve the unity of Iraq against Kurdish separatism—a unity to which all countries, including the United States, are committed.

- Turkey, whose acquiescence is vital to further U.S. military operations, would be ambivalent about such a move by Saddam against the autonomous Kurdish region.

Any such military action by Saddam would immediately place U.S. policy in an extremely difficult situation. U.S. forces are formally committed only to the defense of the protected zone; new policy choices would have to be made about defending the no-fly zone if no Iraqi aircraft were involved. Beyond that however, the United States has a broader but ill-defined moral commitment to protect the Kurds from further genocide at the hands of Saddam. The United States would probably be forced to use air power against Saddam's ground forces invading the Kurdish region, but the decision would involve new commitments.

It is conceivable that a frustrated and enraged Saddam could attempt terrorist acts against the United States. Such a possibility has long existed, however, but even during the Gulf war did not come about. Iraqi terrorists have not even struck against U.S. installations or personnel abroad—in regions far more accessible than within the United States itself.

The Turkish Factor

Allied commitment to the protection of the Kurdish zone in northern Iraq is vastly complicated by the struggle of the Turkish Kurds for greater autonomy within Turkey. This movement is led by a radical, dictatorial, and violent movement, the Kurdish Workers Party (PKK), which conducts a guerrilla war against Turkish authorities from outside the country. Turkey is now expending a massive and costly military effort to crush the movement and in the process is alienating most of the Kurdish population in Turkey, even if much of that population does not sympathize with the PKK in principle. In the process of this brutal civil conflict, Turkey has drawn the disapprobation of its NATO allies, the United States, and the European Community for its violation of human rights in pursuit of a military solution to a political problem. This fact in itself is weakening Turkey's prestige, economic situation, and domestic tranquillity.

If Saddam is to move against northern Iraq to regain control, the campaign will place even greater strain upon Turkey. Turkey will once again have to grant the allies freedom of military movement to operate against Saddam—a divisive political issue within Turkey and even more so in the Arab world. Large numbers of Kurdish refugees from Iraq would almost surely once again flee into Iraq, further exacerbating the domestic situation inside Turkey. While Turkey is committed in principle to Western policies on Iraq, its general staff is increasingly convinced that the existence of an autonomous Kurdish entity in northern Iraq is helping fuel Kurdish separatism within Turkey itself. This judgment is debatable. There is probably little doubt that the creation of an autonomous region there—not originally intended by any player—has probably accelerated the development of political consciousness of Turkey's Kurds and provided a nearby model of an autonomous status. On the other hand, that consciousness was bound to grow as a result of the general growth of ethnic awareness and the search for greater cultural expression among the world's minorities.

Turkey has grown increasingly restive about the continued applications of the sanctions against Saddam Hussein as well, since Turkey's loss of revenues resulting from the closure of the Turkey-Iraqi oil pipeline has cost Turkey perhaps up to $15 billion dollars in revenues over the past three years. As a result, Turkey is seeking for ways to bring the UN sanctions to an end, partially justifying this action by a belief that the sanctions are not working and that it is time to recognize reality and deal with Saddam—albeit much weakened. The French and the Russians are also moving in this direction. The United States will be unlikely to be able to maintain the UN sanctions much beyond the summer of 1995 unless Saddam commits some blunder—a quite conceivable eventuality. At the same time, Turkey believes that the autonomous status of the Kurds of northern

Iraq threatens the creation of a permanent and independent break-away Kurdish state there. Ankara, therefore, believes that the Iraqi Kurds must reach political accommodation with Baghdad—even while Turkey claims it is committed to protecting them against military attack, oppression, or acts of vengeance from Saddam. The mechanisms by which this protection would be guaranteed by Ankara, however, are not clear.

As a result of these many factors, the crisis with Saddam may be reaching a critical juncture. A race is under way between his own fatal weakening and collapse within Baghdad and the weakening and collapse of the sanctions regime imposed from outside. The longer the situation continues, the more unhappy Turkey becomes with the process. At the same time, Turkey's domestic problem with its own Kurds is costing the country more deeply, both in economic and international political terms. U.S. policy toward Ankara is thus rendered more complex, with some short-term conflict of interest.

In broader terms, a basic geopolitical shift has taken place in the region involving Turkey. It is clear under any circumstances that Turkey will be permanently more involved in the politics of northern Iraq than ever before. The Kurdish factor will be an ever more salient situation in international politics in the region. U.S. policy will likely face potential continuing requirements for involvement in the protection of the Kurds in northern Iraq until Saddam falls, and the United States may be called upon to protect them from military action by Saddam before then. Turkey's geopolitical role is shifting southward, and could well involve some security support to the Persian Gulf states as well in the next decade, providing a counterweight to Iran and Iraq. Turkey will figure more prominently in U.S. geopolitical thinking in the years ahead, given its new involvement not only in Iraq, but in Balkan, Caucasian, and Central Asian politics. Its position as a key U.S. ally suggests that the United States will be affected by these Turkish actions.

KUWAIT—THE CONTINUING THREAT FROM IRAQ

While the major security concern over Baghdad relates to northern Iraq, Kuwait itself also remains under possible threat. Baghdad has so far conspicuously refused to recognize Kuwait's sovereignty and right to secure borders. There is no reason to believe that Saddam

has given up in any way—except tactically and for the moment—
Iraqi claims on Kuwait. While a renewed attack on Kuwait would
seem virtually suicidal from Saddam's point of view, his own mentality often operates along quite different lines. In desperate straits,
and consumed with feelings of revenge against Kuwait for the outcome of the Gulf War, Saddam could well be inclined to take some
military action against Kuwait as a face-saving device—miscalculat-
ing sharply again about UN reaction. In fact, however, it would be
impossible for the United States to recreate the same response
against Iraq as last time; Saddam might count on the fact that the
considerable diplomatic problems for the UN in attempting to create
a force to dislodge him—from even a partial seizure of territory—to
be sufficient as to give him considerable room for maneuver. This
continuing threat against Kuwait raises the prospect of continuing
attention to U.S. prepositioning of materiel in Kuwait to meet future
threats; its location would obviously need to be secure, however, and
not vulnerable to a lightning Iraqi attack.

CIVIL WAR AND INSTABILITY IN YEMEN

The former states of North and South Yemen, long at ideological log-
gerheads (South Yemen was the only communist regime in the Arab
world), broke apart in May 1994, engaged in a brief but bloody civil
war, and ended with the conquest of breakaway South Yemen by the
North (San'a) in July. While the civil war is over, deep resentments
remain within the south that could flare up again if San'a pursues
policies either vengeful or otherwise prejudicial to southern interests. Thus the situation remains unstable.

Yemen is of strategic importance in the region for several reasons:

* Combined Yemen has a larger population than Saudi Arabia. It
  represents the single biggest potential threat to Saudi Arabia on
  the whole Arabian Peninsula—as does Saudi Arabia to Yemen.
  Yemen has long-standing territorial claims on Saudi Arabia and
  vice-versa. Saudi Arabia, of course, has vast financial resources
  that Yemen lacks.

* Yemen has recently discovered modest oil reserves that can now
  enable the country to develop more rapidly and to build a military capability. Yemen's military has long been involved in war

between the north and south in past decades and therefore has some practical experience in combat. Saudi Arabia has challenged the territory on which much of this oil is located, primarily in an effort to deny it to Yemen.

- Iraq has long had a significant political presence (the Yemeni Ba'th Party) in Yemen. Because of Yemeni-Saudi friction in the past, Yemen supported Saddam Hussein in the Gulf War, causing Saudi Arabia to expel nearly a million Yemeni workers from Saudi Arabia, critical to Yemen for hard currency remittances. Iraqi military advisors have been assisting the north in the civil war against the south.

- The Yemeni civil war thus took on a broader strategic character, involving other fault lines of Arab politics. Saudi Arabia and the other Gulf states, while ostensibly neutral, were de facto strongly supportive of the south. Fundamentalist Sudan supported the north, as did Jordan. The Yemen conflict has thus mirrored some of the splits of the Arab world during the Gulf conflict. The war now seems to be over, but if tensions are not resolved and a longer term struggle between north and south continues, it could easily become a significant proxy war in the region involving many different states—primarily pivoting on the issue of friendliness or hostility to Saudi Arabia.

- Yemen potentially can become a significant center of anti-Saudi politics and activity in the future; Saudi Arabia is nearly always inclined to play hard ball with Yemen rather than to seek any reconciliation. Yemen has a significant diaspora community around the Middle East, especially in the Gulf; Yemenis represent an active and highly enterprising people who could pose problems to the Saudis on a political level. Saudi Arabia itself has a large Yemeni community apart from laborers, many of whom are well-integrated into Saudi society. Many of them are conservatives from the south who fled the communist regime in the 60s and 70s.

- Oman, which also feels threatened by Saudi power, is a likely ally of Yemen over the longer term. Together they surround Saudi Arabia around the entire southern rim of the peninsula and northward several hundred miles up the Red Sea and the Persian Gulf.

- Yemen by itself is unable to form any serious threat to the stability of Saudi Arabia—short of engaging in open guerrilla warfare or terrorism on Saudi soil—as has sometimes happened in the past with isolated bombing incidents. But should Saudi Arabia itself move toward internal instability, external influences from Yemen could serve to exacerbate the situation.

Any resurgence of the Yemeni civil war could potentially broaden to include Saudi Arabia as a covert or overt belligerent. The south is unlikely to be able to wield an independent armed force again, but it could—especially with Saudi covert assistance—develop the capability for a sustained guerrilla war against the north. While there would be no need for U.S. intervention into the situation, it could become one of the major tension points in the Middle East–Persian Gulf area, reflecting one of the basic have versus have-not fault lines of the region.

SUDAN: FUNDAMENTALISM IN POWER

Sudan is important because it represents the first Sunni Islamic fundamentalist government. (Iran, the only other fundamentalist regime in power, is Shi'ite.) Sudan has taken upon itself to spread the mission of Islam in the surrounding region, especially in North Africa and southward into sub-Saharan Africa. The National Islamic Front (NIF) has ties with other fundamentalist (Islamist) parties in the region and has held several large congresses with representatives from those parties to discuss the future of the movement. Despite Sudan's considerable ambitions to be leader of the worldwide Islamist movement, its ambitions vastly outrun its abilities. First, the country is economically in desperate shape, deeply in debt. Second, it has almost no allies in the region and has lost all Arab foreign aid.

Third, Sudan is not in a position to provide meaningful assistance to propagate Islamism in most of the region: Any Sudanese "assistance" to Egypt or Algeria, for example, has no significant impact on what is almost exclusively a domestic problem. More meaningfully, Sudan is trying to strengthen Muslim groups in neighboring Eritrea, Ethiopia, and Uganda, where its impact may be somewhat greater in those states with their large (or dominant) Christian populations.

It is rather Sudan's civil war that creates the greatest international concern. While the campaign of the basically Arab-Muslim north to Islamize and Arabize the ethnically distinct, Christian-animist south has been going on for decades, it has turned into a serious civil war in the past decade and has intensified since 1989 with the accession of the NIF Islamist regime to power. The death toll is very high, with starvation, massive refugee flows, and gross violations of human rights on all sides. The UN has condemned the Islamization campaign of the Sudanese regime, and many states have made efforts to try to bring a halt to the civil war. African states have been busy both serving as intermediaries and trying to unite the split factions of the southern liberation movement to fight more effectively. The United States has also associated itself with this effort.

It is this situation in southern Sudan that is most likely to spark the prospect of eventual international intervention. Action would be taken first of all to cope with the growing starvation and the massive refugee flows, both internal and external. The Sudanese regime has denied that there is a significant starvation problem and does not want international intervention except for food aid, and under highly controlled circumstances. So far the "CNN factor" has not figured in the situation—there is little media attention on the dimensions of this problem in Africa's largest country; were the media to focus on it, however, the problem could rise in its political salience in the West.

A split of Sudan into at least two states is also a growing prospect. While this would not seem to be of major concern to Western states, it is important to remember that Africa so far has largely been immune to major breakaway movements. It is not that ethnic problems do not exist, but rather that their eruption in a continent where almost no borders coincide with ethnic groups would open a Pandora's box of truly disastrous proportions: potentially endless conflict of division and further subdivision in virtually every state on the continent. The ethnic situation in neighboring states to the east, south, and west of Sudan would all be affected by successful breakaway movements in Sudan's south.

U.S. interest would also be more than normally involved in Sudan because of the intense Egyptian interest in the problem—America's closest Arab ally. Egypt is intensely concerned with instability in

Sudan, first and foremost because the lifeline of Egypt, the Nile River, has its longest passage through Sudanese territory. Any hostile control over the Nile waters could put Egypt at mortal risk. A breakaway movement in southern Sudan would place yet another unstable state astride the Nile waters, a situation Egypt finds nearly intolerable. Egypt has therefore decided not to weaken the Khartoum regime's campaign against the south so as to preserve Sudanese unity at all costs—despite Sudan's poor relations with Egypt and its hosting of banned Egyptian Islamist activists in Sudan.

The United States is highly unlikely to take the lead in sending any U.S. troops to Sudan. On the other hand, eventual UN intervention on a humanitarian or peacekeeping basis is quite a distinct possibility, given the salience of the issue and the dimensions of the problem which probably surpass those of Somalia. The United States could well be involved, at least in a logistics role. While Sudan is a far more developed state than Somalia—at least in the north—a breakdown of order and administration in southern Sudan could potentially pose state-building tasks similar to those that typified Somalia.

THE PERSIAN GULF: IRANIAN THREATS?

Iran was one of the chief beneficiaries of the Gulf War, since its most dangerous enemy, Iraq, was vanquished and placed under a regime of international surveillance for the long term. In principle, Iran has been in a superb position to strengthen its diplomatic position in the Gulf since the end of the war by improving its relations with the small Gulf states in a shared position of opposition to the Iraqi threat. In fact, however, Iran has failed to do so. Iranian policy has been internally contradictory, leading it to rhetorically threaten some of the shaykhdoms on occasion, and to strengthen its administrative and military hold on the island of Abu Musa, whose ownership has been disputed in a volatile fashion with the United Arab Emirates ever since the 1970s under the Shah. As a result, Iran continues to enjoy the distrust of nearly all shaykhdoms, even while they would welcome Iran as a counterweight to Saddam's Iraq.

Iran has also disturbed the Gulf states by its continuing buildup of its military forces. In fact, the build-up is not altogether surprising considering Iran's vast military depletion during the Iran-Iraq war. It is not so much the quantities of Iranian weapons—not unusual for a country of Iran's size and not representing a disproportionate share of its budget—but the types of naval weapons in particular that could affect the Gulf states'—and, indeed, non-Gulf, especially American— maritime or naval operations in the Gulf. Submarines purchased from Russia, and antiship missiles are among the most significant. Iran also has nuclear ambitions, even though its ability to develop such capabilities is still considered by most experts as quite distant.

The Gulf shaykhdoms will remain permanently concerned about Iranian intentions in the region, particularly as long as that state is dominated by a radical Islamist regime, which in the past has called for the overthrow of "corrupt" or "non-Islamic" regimes in the area. Yet the period of the pro-Western Shah of Iran in the 70s in fact represented one of the most military expansionist regimes in the modern history of Iran. Iran will remain the primary fact of life in the Gulf for the remaining states. Iran's military prowess, however, obviously is very unlikely to involve any land invasion of those states. Iran has no access to any Gulf state without first crossing Iraqi territory. Iran, however, is likely to attain the premier naval capabilities among all the Gulf states over time, given its size and the length of its Gulf coastline. Iran's destabilizing subversive and propaganda capabilities are at least as worrying to local regimes, especially as they themselves begin to face greater pressure from their populations for liberalization.

Recently, U.S.-Iran relations worsened further as the United States imposed a unilateral ban on all exports and imports from Iran because of such factors as Iranian opposition to the Middle East Peace process and the alleged Iranian nuclear program. The United States will need to remain alert to Iranian military intentions in the Gulf, dominated primarily by the potential contestation of other islands, off-shore oil wells, and domination of the shipping lanes at the straits of Hormuz.

CONCLUSION

A number of situations in the Middle East today, then, pose significant political problems to U.S. interests and those of its allies. Potential U.S. military involvement in some capacity cannot be ruled out in Algeria, Yemen, Turkey-Iraqi Kurdistan, Kuwait, the Gulf, and Sudan. Proliferation problems will be a permanent concern in Iraq,

Iran, and eventually Algeria, at the least. U.S. interests are particularly complicated by the fact that the major threat to most U.S. allies will involve internal rather than external threats. These are challenges about which the United States can do little if anything except on the diplomatic level in offering council—usually not welcomed. Internal political struggles that are bound to increase over time rather than decrease as these states move toward facing long-overdue political, economic, and social changes.

Chapter Ten

CAUCASUS AND CENTRAL ASIA

Paul Henze

INTRODUCTION AND BACKGROUND—THE CAUCASUS

The southern borders of the ex-Soviet Union remained essentially unchanged from those of the Russian Empire before it. In both the Caucasus and Central Asia, these borders served as the interface between Russian power and the Islamic world. They were the result of a steady Russian southward drive that began in the mid-16th century with the conquest of the Tatar khanates of Astrakhan and Kazan by Ivan the Terrible. Russia continued to make territorial gains at the expense of the Ottoman and Persian empires through the first three decades of the 19th century.[1] During the remainder of the 19th century through World War I, Russia and the Ottomans fought several wars along the Turco-Caucasian frontier. By the Treaty of Berlin in 1878, Russia gained Kars, Ardahan, and Batumi from

---

[1]While Russia had consolidated its position along the Turkish and Iranian frontiers by 1830, its armies were unable to gain full control of the mountain regions to their rear and had to rely on a single major land route over the main Caucasus range—the Georgian Military Highway through the Daryal Gorge. In both Dagestan and Chechnia on the east, and in the Circassian lands to the west, guerrilla movements motivated in large part by Islam long defied both the Tsar's armies and frustrated Russia's divide-and-rule efforts. The Caucasian Mountaineers were not officially "pacified" until 1864. Nevertheless, they continued to revolt whenever Russia was diverted or its power seemed to be weakening. Like the three major peoples of Transcaucasia, the Mountaineers declared independence at the end of World War I and were even supported for a short time by the Bolsheviks. Russia officially observed the 130th anniversary of "the ending of the Caucasian War" on May 22, 1994, with ceremonies in Moscow and in several republican capitals in the Caucasus.

Turkey. Russia's exploitation of the Armenian minority in eastern Anatolia contributed to the catastrophe that befell this unfortunate people during World War I. The ethnically Georgian population of northeastern Turkey, which had been completely converted to Islam by the 18th century, remained in place.[2] Soviet relations with the new Turkish Republic were initially friendly but cooled quickly and remained proper but formal until Stalin's demands on Turkey at the end of World War II drove the country into the developing Western alliance.[3]

Persia's cession of northern Azerbaijan (which then included territory that now forms part of Armenia) to Russia in 1828 marked the end of Russian territorial gains on the Caucasian-Iranian frontier. Henceforth, Russian ambitions toward Persia were characterized by continual maneuvering for economic and political influence and, whenever geopolitical circumstances were favorable, military intervention.[4] Moscow backed a Soviet-style republic in the northern Iranian province of Gilan in 1920 but withdrew the following year and shifted its support to the nationalist leadership in Tehran. During World War II, Moscow abetted creation of leftist Azeri and Kurdish republics in northwestern Iran but was forced by U.S.-initiated UN action to withdraw and abandon the adventure in early 1946.

---

[2]The Soviets gave up claim to Kars and Ardahan when they made peace with Ataturk, but retained Batumi and the surrounding region of Ajaria. Although predominantly Muslim, Ajars are Georgian in language and culture. Ajaria became an autonomous republic within Sovietized Georgia. Benefiting from trade with Turkey since the late 1980s and efficiently administered by a local strongman, Arslan Abashidze, it now enjoys peace and relative prosperity in strife-torn Georgia.

[3]Mustafa Kemal (Ataturk) and his close associates were genuine admirers of Western constitutional government and never found communism, as such, appealing. Early communist efforts to infiltrate the Turkish revolution brought the brief period of opportunistic cooperation with the Soviets to a rapid end. See Harris (1967). Stalin's demands included surrender territory in the east and joint administration of the Turkish Straits.

[4]Britain was Russia's main contender for influence in Iran until after World War II, when the United States inherited that role.

INTRODUCTION AND BACKGROUND—CENTRAL ASIA

Russia took control over portions of the Kazakh steppes in the 18th century but did not begin serious military movement into the Central Asian heartland (Turkestan) until the mid-19th century. Here, Russia came up against the influence and aspirations of China and Britain, activating the "Great Game" that steadily gained momentum from the 1840s to the beginning of the 20th century.[5] While Russia consolidated control over Tashkent in 1867 and a few years later conquered and eliminated the Khanate of Kokand, it let the khanates of Bukhara and Khiva remain as protectorates. The strongest resistance the Russians encountered in Central Asia was that of the Turkmen tribes, which were finally subdued only in 1884.

While Russia was taking control of Turkestan and Britain was consolidating its hold on Kashmir and the northwest frontier of its Indian Empire, Afghanistan survived as a buffer between areas of Russian and British predominance because neither imperial power, in face of fear of the other, was willing to expend the resources necessary to subdue it. Erratic British and Russian efforts to gain influence over an independent warlord in East Turkestan, Yakub Beg, led to the frustration of both at the end of the 1870s.[6] Where Russian, British, and Chinese aspirations all came up against each other—in the Pamirs, the "Roof of Asia"—the Wakhan Corridor was left as part of Afghanistan. Boundaries that still remain official at the end of the 20th century were finally delineated only in 1907.

The Soviets inherited and maintained Russian imperial boundaries throughout Central Asia but detached Outer Mongolia from China and set it up as the first communist satellite in the early 1920s. They maintained Uriangkhai as the pseudo-independent republic of Tannu Tuva until 1944 and then absorbed it as the "autonomous" republic of Tuva. During the previous decade, taking advantage of

---

[5]The most comprehensive account is Hopkirk (1990).

[6]Yakub Beg was actually a native of the Khanate of Kokand, i.e., territory that the Russians conquered in the 1870s. I recounted this colorful episode in the history of Chinese Turkestan in "The Great Game in Kashgaria" (Henze, 1989). China renamed the region Xinjiang, the "New Dominion," after reconquering Yakub Beg's "Independent Kashgaria" in 1878. After conquering it in 1949, the Chinese Communists designated it the Xinjiang-Uigur Autonomous Region in 1955, giving recognition to its major indigenous nationality, the Turkic Uigurs.

China's confused political condition, they had for all practical purposes turned Xinjiang (East Turkestan) into a colony but refrained from formally incorporating it.

In the early 1920s, as soon as the Red Army had gained control of the main urban centers and lines of communication in Central Asia, the Bolsheviks abolished the Khanates of Bukhara and Khiva, as well as the Governorate General of Turkestan, and restructured the region along somewhat artificial ethnolinguistic lines.[7] The Bolsheviks were motivated by a combination of fears—Islam, Turkestani nationalism, and anticolonial sentiment. The region was gerrymandered into five separate republics, all of which had been "raised" to union level— theoretically coequal with the Russian Federation as components of the Soviet Union—by 1936.[8] Strenuous efforts to obliterate Islam and create a strong sense of secular identity among members of each newly defined Central Asian ethnic group continued through most of the Soviet period with mixed results. Doubts about the extent to which the rapidly increasing population of Soviet Central Asia had actually been converted to communism played a large part in the decision to invade Afghanistan in 1979, the reverberations of which are still being felt in Tajikistan and Uzbekistan. Well before the collapse of Soviet power, it had become apparent that Islam, especially in its cultural dimension, remained an inherent feature of the identity of most indigenous Central Asians.

Since the Soviet collapse, every Central Asian republic has experienced a resurgence of Islam, but there has been little sympathy for Iranian-style fundamentalism. The degree to which these peoples may be susceptible to the appeals of other forms of Islamic radicalism remains to be seen. Both Russians and some Central Asian communist-successor leaders have exaggerated the danger for their

---

[7]The sense of ethnicity was not highly developed in Central Asia. The socioeconomic distinction between settled agriculturalists and pastoralists was more significant in much of the region. Except for small minorities of Jews and Ismailis, almost all indigenous Central Asians were adherents of Sunni Islam. Turkic and Persian speakers were intermixed.

[8]Tajikistan, originally an autonomous republic within Uzbekistan, was made a union republic in 1929. Kazakhstan and Kyrgyzstan were separated from the Russian Federation and made union republics in 1936. Use of the term Turkestan was discouraged and ceased officially. It survived, curiously, only in the designation of the Turkestan Military District, which retained this title during the entire Soviet period.

own purposes. China abandoned anti-Islamic policies after the death of Mao, with the result that religion is flourishing among its Muslims, although there is little evidence of Islamic radicalism. This experience bodes well for ex-Soviet Central Asia.

Soviet policies were more successful in fostering ethnic nationalism in Central Asia than in suppressing religion. On becoming independent in 1991, each republic accepted its existing borders and took measures to strengthen its ethnolinguistic identity. Problems of definition between Uzbeks and Tajiks, however, are a factor in the strife that has plagued independent Tajikistan, where the effects of continued instability in Afghanistan are also felt. The apparent interethnic hostility that led to disturbances and bloody riots during the final years of Soviet control in all Central Asian republics except Turkmenistan reflected economic and social strains as much as ethnic tensions *per se*. Among intellectuals, at least, the concept of Turkestan as a single cultural-political entity is far from dead. The historic cultural and geographic unity of the region has encouraged practical arrangements for economic and political cooperation.

THE BURDEN OF HISTORY

The collapse of the Soviet Union did not leave a *tabula rasa* in either the Caucasus or Central Asia from which political life could begin anew in pristine form. Instead, it cut these societies loose from tight control from Moscow and in varying degrees opened them up for expression of long-suppressed concerns that could not be acknowledged publicly in the Soviet period.[9] As a result, history has come alive with a vengeance after lying dormant for 70 years. For the peoples of the Caucasus and Central Asia, history matters to a degree difficult for those who live in the open societies of the Free World, or even in many other excolonial countries, to grasp. Ethnic rivalries, territorial claims, and demands for rectification of past grievances dominate political life in most of these countries. In some (notably

---

[9]In some of these countries, e.g., Georgia and Azerbaijan, leaders who initially enjoyed overwhelming public support and promised reforms lacked political skills and quickly provoked opposition. They exacerbated economic, political, and social deterioration, which has fed on itself. In others, e.g., Uzbekistan and Turkmenistan, communist leaders transformed their parties into neoauthoritarian oligarchies and continued to govern in much the same fashion they had followed as Kremlin satraps.

Armenia), such preoccupations have severely constrained prospects for economic and political reconstruction. Therefore, both the causes and the likely course of developments that affect the security of these regions and impact on the international community must be analyzed in a historical context.

THE SIGNIFICANCE OF RUSSIA

In spite of the collapse of the Soviet Union in 1991, Russia looms large in the awareness of the leaders and the peoples of all the Caucasian and Central Asian states. In general, Russia has the least influence—directly—on their politics, much more on their economic situation, and most on their security. In none of these countries is Russia's influence exercised consistently, for Russia itself is in a highly transitory condition with competing power structures. Processes of decentralization are likely to accelerate for at least the remainder of the 20th century. *De facto* fragmentation could be the result. Institutionally, the Russian state has not yet gained a clear identity separate from what was left over from the Soviet Union. In many parts of Russia, the bases for democracy at the grass-roots level have not yet been created. At the national level, political parties and political processes continue to operate chaotically. This makes Russian behavior vis-à-vis the newly independent ex-Soviet states difficult to calculate and makes their leaders and peoples uneasy about Russia's actions toward them. The muddled and bloody Russian military intervention in Chechnya at the end of 1994 has compounded these problems.

The situation is exacerbated by the fact that Moscow has been unable to enunciate or enforce clear policies toward the independent states that form its southern periphery, or even toward neighboring ethnic and regional components of the Russian federation.[10] The

---

[10]The years 1994 and 1995 brought a resurgence of declarative Russian nationalism among politicians and increased rhetorical assertiveness by Russian military leaders toward the entire "Near Abroad," combined with partially successful efforts to pressure the Transcaucasian governments to accept Russian military bases. The Chechen misadventure can be seen on the one hand as a manifestation of the nationalistic aggressiveness of portions of the Russian military leadership, as well as other elements in the power structure, such as the only minimally reformed KGB. On the other hand, it has demonstrated striking weakness and political and military ineptitude on the part

invasion of Chechnya demonstrated the extent to which rhetoric, reality, and Russia's capacity for decisive action are often sharply at odds. Different groups in Russia have poorly formulated and often contradictory perceptions of how Russia's national interests are best served. Limited financial resources and atrophying of military forces are a severe constraint on military assertiveness. The Russian public is reluctant to bear the costs of imperial reassertion. Few regions of Russia remit revenues in full to Moscow, and draft calls are seldom met. Many Russian regions are only loosely controlled by Moscow.

The North Caucasus illustrates this problem strikingly, for the degree to which the contested military occupation of Chechnya will result in actual enhanced control remains to be seen. The seven ethnic republics of the North Caucasus,[11] the Kalmyk Republic, the Krasnodar and Stavropol Krays, and the three independent Transcaucasian states (Georgia, Armenia, and Azerbaijan) form a region of distinctive geopolitical and economic interaction. The announcement by the Ministry of Defense in Moscow late June 1994 that a new North Caucasian Military District, encompassing this whole region, was being formed constituted recognition of this fact, but the assault on Chechnya at the end of the year was hardly evidence of a more coherent Russian approach to the region.[12]

From the 17th century onward, Russia's political approach to the ethnic and religious diversity of the Caucasus was *divide et impera* (divide and rule). Camouflaged as *druzhba narodov* (friendship of peoples), this policy was continued, often in particularly crass forms, during the Soviet period. Nationalities were played off against each

---

of all elements involved. It has also provoked widespread criticism in many segments of Russian society. The full consequences of the invasion of Chechnya on other non-Russian components of the Russian Federation are not yet calculable, but it has caused restiveness in neighboring Caucasian republics as well as the independent ex-Soviet states of the Caucasus and Central Asia.

[11]From west to east: Adygeia, Karachay-Cherkessia, Kabardino-Balkaria, North Ossetia, Ingushetia, Chechnya, and Dagestan. The last-named contains ten major and more than a dozen minor nationalities.

[12]Far from representing a creative or promising solution to the challenges the entire Caucasus represents for Russia, the new military district appears to be a reversion to the 19th century policy of dealing with Caucasian restiveness and resistance by military means. It is doubtful that military force can provide any permanent solution to the problems Russia faces along this portion of its southern border; it could, in fact, exacerbate them.

other, and boundaries were frequently juggled.[13] The result since the collapse of communism is continual strain among ethnic groups and actual or threatened violence. The most persistent and destructive violence has been Armenia's war against Azerbaijan over its largely Armenian-populated enclave of Nagorno-Karabakh. Moscow is accused by Azerbaijan of having tilted toward Armenia at key junctures over the past six years. In Abkhazia, while Yeltsin remained passive, various Russian elements, for varying reasons, joined to abet the revolt of the Abkhaz *nomenklatura* against Georgia.[14] Russian military intervention in Chechnya at the end of 1994, which was preceded by several months of thinly veiled clandestine operations against the Chechen leadership, marked a new stage in Russian action in the region.

Most leaders of the newly independent countries of the Transcaucasus and Central Asia task Russia with lack of understanding and/or neglect of their economic problems. The manner in which Russia mismanaged the ruble in 1993 forced most of them to introduce their own currencies, often with little time for careful preparation of the shift. All have suffered financial dislocation as a result, although some of the new currencies (notably that of Kyrgyzstan) have shown greater strength than the ruble. Between the late summer of 1994 and the spring of 1995, the ruble lost more than half of its value. When of necessity—lack of practical alternatives—these countries have reached agreement with Russia on fuel supply and other economic arrangements, they have usually been left with the feeling that Russia has dealt with them cavalierly. At the same time, both official and private Russian reactions to these countries' efforts to diversify their economic relations by expanding trade with Turkey, Iran, Pakistan, India, and especially China and other Far Eastern countries have created resentment and have fed suspicions that Russia aims to reassert economic hegemony over them.

---

[13]The boundaries of the North Ossetian Republic were changed more than 30 times during the 70 years of Soviet rule.

[14]For old communists and conservative Russians, the desire to punish Eduard Shevardnadze for his support, then abandonment, of Gorbachev and consequent contribution to the collapse of the Soviet Union was a strong factor motivating intervention in Abkhazia.

While, legally, Russia's position toward the newly independent ex-Soviet countries of the Caucasus and Central Asia has been consistently proper and the rancor and threats over troop withdrawal and citizenship status that have bedeviled Russian-Baltic relations have been generally absent, mounting neo-imperialist rhetoric in Moscow has provoked alarm in all the Caucasian and Central Asian capitals. Zhirinovsky is at the extreme end of a spectrum that extends back to much more rational and probably more serious proponents of consolidation of the still vaguely constituted Commonwealth of Independent States as, in effect, a resuscitated Soviet Union. The Russian desire to have troops in ex-Soviet countries placed under the official umbrella of the UN as international peacekeepers has provoked opposition among many Caucasians and Central Asians. They see this and Russian desires to share policing of borders as providing a convenient excuse for military-backed Russian interference in their affairs. In 1992 and 1993, Georgia and Tajikistan fell into such disorder that leaders saw no alternative to accepting Russian troops to maintain order. In both situations, there is evidence that Russians—whether operating autonomously or at specific Moscow orders—were major contributors to the disorder the soldiers were then sent to counter.

Russian military intervention in Georgia and Tajikistan has reduced violence in those countries, but it has not eliminated it and has not stabilized deeply troubled political situations. Outright military attack on Chechnya shows little promise of improving the basic situation there or in the North Caucasus as a whole. The Ingush, favored by Moscow over the Chechens in 1991 and 1992, appear to have shifted back to an anti-Moscow position because of Moscow's failure to satisfy Ingush grievances against the Ossetes. Moscow has so far avoided serious difficulties in Dagestan by permitting large sections of this deeply Islamic republic to go their own way. It has avoided sending in military forces. Lezgin demands on Azerbaijan remain a tempting device for Moscow to use to induce Azeri leaders to make concessions over control of petroleum. The Abkhaz problem still bedevils Moscow's relations with Georgia, and no simple solution is in sight.

Too few Russians still live in the Transcaucasian countries to constitute a serious problem, although they could offer an excuse for direct intervention. The notion of intervening to protect them has been discredited by the fact that Russians in Chechnya probably suffered to a greater extent from Moscow's brutal military tactics than the Chechens themselves. Official and unofficial declarations of concern in Moscow about Russians living in the "Near Abroad" are a major source of disquiet for Central Asian leaders in Kazakhstan and Kyrgyzstan.[15]

Russian leaders must be concerned about maintaining control over their own vast federation as it continues to go through a process of economic and political transformation. Russia is troubled not only by ethnic assertiveness among sizable groups of non-Russians but also by the reluctance of many important regions (the Northwest, the Urals, Central Siberia, Sakha-Yakutia, the Far East) to accept a high degree of Moscow management of their affairs.[16] Russian leaders fear collaboration between some of these ethnic groups and regions and the newly independent former Soviet states beyond Russia's borders and in some instances even beyond the old Soviet borders (in the case of the Mongols, for example).

INTERNATIONAL RESPONSE TO INDEPENDENCE

The break-up of the Soviet Union was quickly recognized by the international community. The newly independent states of the Transcaucasus and Central Asia were admitted to the UN and subsequently joined many other international organizations and groupings. With little delay, the United States and most U.S. allies established diplomatic relations with these countries. The new nations joined the World Bank and the International Monetary Fund, as well as regional economic groups. Many countries, including the United States, inaugurated emergency relief and economic assistance programs. Foundations, endowments, and government agencies (including the Peace Corps) have inaugurated political training, educational, and cultural programs. Without exception, the newly inde-

---

[15]While we continue to hear about Russians in Kazakhstan, little attention is paid either in Russia or abroad to the fact that at least 700,000 Kazakhs live in Russian regions directly northwest of Kazakhstan.

[16]Tatars, with a total population over five million, are the most numerous ethnic group in the Russian Federation after the Russians themselves. Tatarstan recently concluded a treaty with Moscow under which the republic retains full control over its internal administration.

pendent states were eager to have their independence recognized, to send representatives to international bodies, to become eligible for economic assistance, and in varying degrees, to welcome private investment. All committed themselves to establishment of democracy and open societies and accepted membership in the Organization for Security and Cooperation in Europe (OSCE). Since their governments were all inexperienced in conduct of foreign and international relations and since several of the new governments turned out to be comparatively unstable, the performance of these states has varied.

Turkey and Iran immediately showed a strong interest in the newly independent Transcaucasian and Central Asian states. This interest was for the most part reciprocated—warmly in the case of Turkey, more cautiously in respect to Iran. The Central Asian states also took rapid measures to explore economic relations with China, Japan, South Korea, Taiwan, and other countries of East and South Asia. As time has passed, the relations of the newly independent states with all of these countries have developed along more differentiated lines. As far as Turkey is concerned, excessively high expectations on both sides have become dampened over the past two and a half years, and a condition of healthy balance has for the most part resulted. Iranians who thought Islam could be exploited to forge both direct and indirect influence in Central Asia have been disappointed, while, at the same time, Central Asian leaders who feared Iranian initiatives have been gratified to find that their populations have responded skeptically to Iranian overtures. This has proved true even in Tajikistan. Chinese trade with Central Asia has expanded greatly. Central Asian trade with Pakistan and India has also greatly increased.

Only one of the new states had nuclear weapons on its territory: Kazakhstan. Kazakh leaders were initially inclined to disavow any interest in retention of them. Western concern about them—demands that Kazakhstan commit itself to turn over all nuclear weapons to Russia—alerted the Kazakh leadership to the political value of these weapons and complicated the process of gaining a commitment for their removal and of assisting and compensating Kazakhstan for becoming a nuclear-free country. Kazakhstan finally adhered to the Nuclear Non-Proliferation Treaty in December 1993, and signed an agreement in July 1994 with the International Atomic

Energy Agency (IAEA) on monitoring of its commitment to dismantle and destroy all nuclear weapons and delivery systems on its soil. This was accomplished during the first half of 1995. In return, the IAEA agreed to assist Kazakhstan in coping with pollution from Soviet-era testing and related activity. The nuclear issue is thus no longer a contentious factor in Kazakhstan's relations with the West.[17]

While tactical nuclear weapons may well have been deployed in the Transcaucasus by the Soviet Union, they appear all to have been withdrawn by the time of collapse of Soviet power.

Private U.S. and other Western economic interest in Central Asia has been centered on the region's oil, gas, and mineral resources. Azerbaijan's oil resources have likewise attracted the interest of Western investors. Initial expectations of quick, lucrative deals have proved unrealistic, however, because securing final agreements on concessions and joint ventures has been more difficult than either side anticipated. These are gradually being worked out, but Russia's relationship to all these issues, it is now recognized, must also be taken into account. The most difficult problem that remains to be overcome is transport, because almost all existing routes out of the region to world markets pass through Russia. New pipelines through Iran and/or Turkey or under the Caspian will require a large investment that can only be obtained if a satisfactory degree of political viability over time can be assured. Existing pipelines in Georgia offer the possibility of transport of Azerbaijani oil directly to Turkey. The Russian pipeline route through Chechnya entails high risk of disruption as long as the situation there remains unsettled.

U.S. INTERESTS IN THE CAUCASUS AND CENTRAL ASIA

In its most general sense, the American interest in the newly independent countries of the Transcaucasus and Central Asia is a continuation of the historic anticolonialist policy the United States has fol-

---

[17]According to the CIS Nuclear Database of the Monterey Institute of International Studies as of May 1, 1994, all nuclear weapons in Kazakhstan remained under Russian control at that time. All SS-18 nuclear warheads were then scheduled to be transferred to Russia by mid-1995 and all missile silos dismantled by mid-1997. All strategic bombers in Kazakhstan have been moved to Russia. Associated air-launched cruise missiles remained in storage at Semipalatinsk until early 1995.

lowed since 1776. Following World War I, the United States welcomed the new states that emerged from the Austro-Hungarian and Ottoman empires in Eastern Europe and eventually in the Middle East, where World War II brought the process to a conclusion. After the United States granted independence to the Philippines in 1946, a great wave of decolonization of the British, French, Belgian, and (eventually) Portuguese empires brought dozens of new members, beginning with India and Pakistan, into the UN. Seen from the perspective of 20th century world history, the collapse of the Soviet Union has marked the near-final stage of the break-up of empires and extension of self-determination for peoples living under them.[18]

In immediate pragmatic terms, the United States supports the independence of these states to insure against the revival of imperialist authoritarianism in Russia. A truly democratic Russia, responsive to the needs of its own citizens, is unlikely to accept the costs and risks of trying to revive the Soviet imperial system. The Russian-Soviet Empire can be reconstituted only by authoritarian leaders hostile not only to the "Near Abroad" but to all other neighboring states— Turkey, Iran, China. Under such circumstances Russia could hardly serve as a partner of the United States and its allies in maintaining peace and order, encouraging free trade, and creating conditions for constructive economic and social development for all races and peoples.

Nevertheless, Russia has both strong historic links and legitimate economic and cultural interests in the countries of Transcaucasia and Central Asia.[19] These include an interest in the evolution of

---

[18]The only remaining major untransformed empire is China. Consideration of the course of developments there as communist control weakens further is beyond the scope of this essay. Russia retains imperial characteristics, especially in the North Caucasus, where peoples who regard themselves as conquered and colonized have become politically assertive since the collapse of Soviet power.

[19]Russia is entitled to participate in development of the oil and gas resources of Azerbaijan, Kazakhstan, and Turkmenistan on the same basis as other countries. These countries are naturally opposed to the apparent desires of some Russians to exercise monopoly control over export and marketing of the oil and gas these countries produce. Russian claims to ownership of oil under the Caspian are unacceptable to other littoral states (Azerbaijan, Kazakhstan, Turkmenistan, and Iran), and there is no justification for a Russian veto over investment by Western oil companies in the Caspian. Territorial jurisdiction in the Caspian remains a contentious issue, however.

stable political systems in these countries that can guarantee predictable conditions of life for all their peoples, including Russians
who choose to remain in them. American policy cannot ignore legitimate Russian interests and would be unwise to encourage ex-Soviet
states to maintain a stance of undifferentiated hostility to Russia.
But neither can the United States accept the notion that history enti-
tles Russia to impose privileged dominance in these regions any
more than Britain is entitled to dominance in Pakistan or India, or
France in Algeria or Tunisia.

U.S. interests in the countries of Transcaucasia and Central Asia
include the *positive* aims of

1. Ensuring continued *access* to these societies to encourage democratic political development and constructive exchanges at all
   levels between citizens and institutions

2. Gaining maximum opportunities for mutually beneficial *trade and
   investment* by encouraging rapid evolution of sound *free-market
   economies*[20]

3. Assisting these countries to *develop constructive international cultural, political, and economic relations,* particularly with their
   neighbors, including participation in regional security initiatives.

The *negative* aims of U.S. policy are for the most part the other side
of the same coin. They include the following:

1. *Prevention of economic, social, and political degeneration,* which
   would require major humanitarian aid to avert famine and disease

2. *Prevention of civil war or military action* across borders by restrictions on proliferation of arms and armed forces

3. *Removal of nuclear weapons,* as well as development and testing
   facilities, and control of illicit traffic in nuclear materials

---

The situation in respect to other closed or semiclosed bodies of water offers prece-
dents for regulation.

[20]If substantial American investments materialize in oil, gas, and other minerals in
Kazakhstan, Azerbaijan, and/or Turkmenistan and other Caucasian or Central Asian
countries, the economic significance of these countries to the United States will
increase.

4. Maintenance of *effective controls* over production of and traffic in *narcotics,* as well as other *criminal activities*

5. *Opposition to radical and destructive political and social movements,* whether religious or secular.

Since there is no basis for the United States to claim or justify (in terms of its own national interests) priority in relations with the Transcaucasus and Central Asia, all American policy aims need to be pursued in cooperation with international organizations and other interested countries, including Russia and China, as long as those countries' behavior toward the newly independent states meets recognized international standards. Turkey will continue to take a special interest in the Caucasus and Central Asia. Turkey can also serve as one of several channels through which Iran can hopefully be persuaded to pursue constructive policies toward these countries, deemphasizing religious fanaticism and anti-Westernism.

CURRENT REALITIES AND POLICY CHALLENGES— CENTRAL ASIA

While the principles that guide American policy toward the Transcaucasus and Central Asia seem clear, the rapidity with which these countries became independent has created problems and critical situations that their leaders and governments have not been able to master. Policy formulation and implementation must take these into account. Except for Kyrgyzstan, all Central Asian governments are derived directly from the previous communist ruling groups of their republics. Old authoritarian habits, a sense of insecurity, fear of disorder, and concern about the impact of independent politicians and media on their populations are characteristic in some degree of all of these governments. Leaders have not yet become accustomed to governing with open political opposition, and most opposition groups do not yet understand how to play a constructive role. Many of them are unrealistic in their demands and inept in their methods. Neither the governments in power nor their oppositions have yet had much success in educating their citizens to develop habits of responsible political participation.

Each leader has dealt with opposition differently. Only Akaev in Kyrgyzstan has permitted open political campaigning and a fair degree of media freedom. In Uzbekistan, Karimov has harshly oppressed political rivals and critical journalists. In Turkmenistan, Niyazov's cult of personality has left little room for political dialogue. In Kazakhstan, where Russians still constitute perhaps as much as 35 percent of the population, Nazarbaev has dealt with political opposition and the media with less offensive and oppressive methods than Karimov. While serious civil disorder has been avoided in all the other Central Asian countries since independence, Tajikistan has been the scene of chronic civil war. The war has been misrepresented as a struggle between secular forces and Islamic fundamentalists. It is more a struggle between regional power elites, some of them old communists, and more democratically inclined elements. While all other Central Asian governments exercise effective control over their territories, perhaps as much as 40 percent of Tajikistan does not recognize the authority of the government in Dushanbe.

Almost everywhere in Central Asia, political dialogue has developed in a relatively narrow urban intellectual context with little involvement of people outside the capital cities. Some foreign criticism of the slowness of democratization and reform has been based on shallow understanding of the nature of Central Asian traditions and the effects of more than a century of Russian imperial and communist rule. Foreign support of opposition politicians has sometimes been counterproductive of the results it aims to achieve. An important lesson of experience to date seems to be that patience and subtlety and a keener sense of Central Asian leaders' perceptions of the formidable problems they face will be more effective in creating participatory societies and responsive governments in these new states than confrontational approaches.

DETERIORATION IN THE CAUCASUS

In the three Transcaucasian countries, leaders who came to power in the first two years of independence, unlike those in Central Asia, all symbolized and desired a sharp break with previous communist power cliques. Only one, Ter-Petrosian in Armenia, has survived in office, although he faces rising opposition and in 1995 resorted to extreme measures to deal with some political groups opposing him.

Armenia

Armenia is an extreme example of preoccupation with grievance. Ter-Petrosian was unable to halt or limit the campaign to wrest Nagorno-Karabakh from Azerbaijan. It turned into a general war against Azerbaijan that has undermined Armenia's own viability. Armenian militants have been strongly influenced and reinforced by the Armenian diaspora, including Armenians from the United States. Economically and socially, Armenia has never recovered from the major earthquake of 1988. The investment required to sustain steady warfare since independence has crippled the economy in spite of relatively generous foreign aid. At least 750,000 citizens—almost a quarter of the population—are believed to have emigrated.

Georgia

Georgia, like Armenia, possesses a highly educated and industrious population and has a stronger economic base but also fell into deep political and economic crisis. Its first president, Gamsakhurdia, exacerbated preexisting strains with Ossetes and Abkhaz and provoked bitter rivalries among Georgians themselves while neglecting economic reform. He was driven from power in the first days of 1992. Former Soviet Foreign Minister Shevardnadze was persuaded to return a few weeks later to try to restore equilibrium. Shevardnadze was unable to restrain Georgian warlords eager to fight in Abkhazia, where the situation worsened steadily.[21] Large quantities of arms and equipment diverted from Russian bases, the participation of North Caucasian and Cossack mercenaries, and direct involvement of Russian military personnel enabled Abkhaz separatists to expel all Georgian forces in several bloody weeks of fighting in the fall of 1993. Some 200,000 to 300,000 refugees fled to territory still controlled by Georgia, while, in supremely

---

[21]Abkhazia, a rich and scenic region along the Black Sea, was the playground of the Soviet communist elite. During the Soviet period, Moscow manipulated the Abkhaz as a counterweight to Georgian nationalism, and the region became an extreme example of the Soviet divide-and-rule strategy toward ethnic groups. Abkhaz accounted for only 17 percent of the population by 1989, but a small Abkhaz *nomenklatura* monopolized most of the lucrative administrative, economic, and cultural positions. Fearing loss of their privileged status in an independent Georgia, these people became natural allies of Russian reactionaries.

Machiavellian fashion, Russia reversed roles, sent marines to rescue Shevardnadze and helped Georgian troops loyal to him defeat an insurrection of Gamsakhkurdia loyalists.[22]

In return for rescue by Russia from the predicament into which Russian support for the Abkhaz had plunged Georgia, Shevardnadze had to agree to join the CIS and consent to continuation of three Russian bases in Georgia.

Shevardnadze's appeal to President Clinton during his Washington visit in March 1994 for American participation in an international peacekeeping force in Abkhazia produced no response.[23] Georgia accepted Russia's offer to provide peacekeepers in Abkhazia at the beginning of the summer of 1994, and the UN agreed to increase its small observer contingent to 150 men.

The situation in South Ossetia has remained unsettled, but a joint Georgian-Russian peacekeeping arrangement agreed to in 1992 has succeeded in preventing further severe outbreaks of fighting. Gradually, during late 1994 and 1995, Shevardnadze succeeded in expanding the Tbilisi government's authority in the rest of Georgia. Its most peaceful region, Ajaraia, remains firmly controlled by Abashidze, who supports Shevarnadze

North Caucasus

North of the main chain of the Caucasus, Russian political mediation and military threats were markedly unsuccessful in reducing or solving conflicts during the three years preceding the invasion of Chechnya. Economic decline and political confusion in both the western republics (Kabardino-Balkaria, Karachai-Cherkessia, and Adygeia) and in ethnically complex Dagestan on the east had caused a steady flow of refugees—both Russian and other nonindigenous

---

[22]Gamsakhurdia took refuge in Chechnya when he fled Tbilisi in January 1992. Returning to lead his supporters in the fall of 1993, he reportedly committed suicide when they were defeated. His body was taken from Georgia to Grozny for reburial in February 1994. The fact that the Chechen leader, Jokhar Dudaev, gave Gamsakhurdia asylum and supplied mercenaries to fight against Georgia in Abkhazia was a major factor influencing Georgia to avoid condemnation of Russian military operations in Chechnya in December 1994.

[23]He cited the precedent of the token U.S. force in Macedonia.

peoples, as well as North Caucasians—to Krasnodar, Stavropol, and more northerly regions of Russia even before the vast outflow of destitute Russians and Chechens that resulted from the invasion of December 1994.[24]

Azerbaijan

The political situation of Azerbaijan, whose oil fields could make it the richest of the three Transcaucasian countries, has parallels with Georgia. The Popular Front government headed by Ebulfez Elchibey elected in June 1992 had no significant accomplishments to its credit when it fell to an internal rebellion widely thought to be abetted by Russians in the summer of 1993. Elchibey had developed close relations with Turkey but alienated Iran by advocating eventual absorption of southern Azerbaijan.[25] This provoked an Iranian tilt toward Armenia. Suspicions of Russian collusion in Elchibey's fall were reinforced when Heidar Aliev, who had served for nearly two decades as Brezhnev's viceroy in Baku, replaced Elchibey.[26] Azerbaijanis who had become disillusioned with Elchibey hoped Aliev would at least be able to reverse what they perceived as a Russian tilt toward Armenia and bring about a settlement of the war.[27]

Though Aliev was initially regarded, especially in Turkey, as a Russian puppet, he has proved to be a relatively astute manipulator of the various forces that come into play in Azerbaijan. He has avoided massive suppression of political opposition. He has bal-

---

[24]The refugee population of the Stavropol and Krasnodar krays had reached at least 600,000 by mid-1994; some sources indicate that it has doubled since then.

[25]Until the Russian conquest of the north during the first quarter of the 19th century, Turkic Azerbaijan was united within the Persian Empire. Consciousness of common nationality has persisted in both north and south. As the Soviet Union began to collapse at the end of the 1980s, Azerbaijanis on both sides of the border dismantled fences and control towers and began to cross freely.

[26]Aliev had been removed by Gorbachev.

[27]In reality, various Russian elements appear at different times to have intervened in support of both the Armenians and the Azerbaijanis. Moscow's role is unclear. In this respect, as in most aspects of Russian policy toward all parts of the Caucasus, Russia has not developed a consistent concept of its own basic interests in the region. Moscow has exercised only intermittent control over regional military commanders and KGB successors. Different elements in Moscow often pursue uncoordinated courses of action and authorize contradictory policies.

anced Russian interests against foreign bidders for oil concessions. Azeri troops began to give a better account of themselves in resisting Armenian advances in late 1993 and even retook some territory, but no real progress has been made toward settlement of the conflict by either Russia or OSCE mediators, and refugees remain a heavy burden on the beleaguered Azerbaijani economy.

THE FUTURE OF AMERICAN POLICY TOWARD THE CAUCASUS AND CENTRAL ASIA

Like the Bush administration before it, the Clinton administration was initially criticized for being oversolicitous of Russia and neglectful of the other Soviet successor states. Following the October 1993 events in Moscow and the December Russian 1993 elections, the first months of 1994 initiated a readjustment in U.S. policy that has continued. The Clinton administration's initial endorsement of the Russian attack on Chechnya provoked strong criticism from many quarters at home and abroad. This led to a modest shift in early 1995 to qualified condemnation of human rights violations and appeals for accelerated efforts to end fighting and find a peaceful solution. As of this writing (summer 1995), the U.S. administration's position shows some degree of reversion to relatively unqualified support of Yeltsin's position.

The basic rationale for American interest in the Caucasus and Central Asia has been discussed in a preceding section. A few additional, specific observations may be useful to put future prospects into perspective.

The United States has already been providing official humanitarian aid to several of these countries. Private American organizations have also been active. Even if cease-fires are implemented in wars in Tajikistan and Armenia-Azerbaijan, the refugee burdens from these and other strife-torn areas (e.g., Abkhazia and Chechnya) are not likely to ease quickly. Major new outbreaks of armed conflict do not seem imminent in the region, but are difficult to predict and depend partially on the lessons the Russian leadership, including military leaders, draws from the experience in Chechnya. It is probable that resistance in Chechnya will continue indefinitely at an intermediate level of intensity. It may spread to neighboring republics. The whole

Caucasus will continue to be highly vulnerable to ethnic strife. Therefore, *humanitarian relief is likely to be needed indefinitely.* Furthermore, many parts of the region are earthquake-prone. Natural disasters could cause sudden emergency requirements.

A share of the economic assistance the United States is committed to provide for the former Soviet Union as a whole is earmarked for the countries of the Transcaucasus and Central Asia. USAID missions have initiated programs in each of these countries.[28] In spite of many difficulties, a substantial number of private American investors see these countries as more hopeful than many parts of Africa, and modest investments have already been made. In time, barring a serious deterioration of political and security conditions, the Transcaucasus and Central Asia could attract substantial American investment. With stabilization of economic conditions, prospects for trade would immediately improve.

Official cultural and educational exchange programs are operating in all these countries. Many quasi-governmental[29] and private American institutions have also initiated projects. These are likely to expand, for demand is great. Both official and private efforts to assist in dealing with environmental pollution are under way. The United States has warmly endorsed the Black Sea Initiative, in which Turkey has taken the lead and will no doubt continue to do so. In several of these countries, assistance is being provided to help local authorities deal with narcotics traffic and other activities of international criminal mafias. These efforts are likely to expand.

Contacts with the newly organized military forces of the Transcaucasian and Central Asian countries have been established. U.S. military and civilian specialists have been assisting Kazakhstan in dismantling nuclear weaponry and advising on clean-up of contaminated areas. The United States does not anticipate supplying military equipment or deploying military personnel beyond attachés and small temporary-duty missions for relief, advisory, or survey

---

[28]The Armenian lobby in Congress has, however, succeeded in maintaining an embargo on developmental assistance for Azerbaijan. In FY 1993, U.S. assistance to Armenia totaled $335 million; to Georgia, $279 million; to Azerbaijan, $34 million.

[29]For example, the National Endowment for Democracy and the Republican and Democratic Institutes.

purposes. A small number of officers from the new countries' armed forces are being given training in U.S. military schools. The administration has expressed a willingness to contribute to costs of peacekeeping missions but is not prepared to provide significant numbers of military personnel. The reluctance of the Republican-dominated U.S. Congress is likely to further inhibit participation in peacekeeping activities. There is less reason to believe that the United States will not continue to be receptive to participation in at least short-term humanitarian relief activities, whether undertaken unilaterally or multilaterally.

Russia has sought to have its forces deployed in the Caucasus and Central Asia designated as UN peacekeepers but without the formal command structure that such operations entail. The United States and most of the international community have been unwilling to see Russian operations in Georgia, Armenia, Azerbaijan, or Tajikistan so designated without participation of forces of other countries and without a separate multilateral command structure. Policy trade-offs with Russia over Bosnia and other parts of ex-Yugoslavia are conceivable. The possibility cannot, however, be entirely excluded that the United States would find itself under pressure (*inter alia* from long-standing allies, such as Turkey and Pakistan) to provide token numbers of military personnel for peacekeeping in the Caucasus and/or Central Asia.

FUTURE MILITARY REQUIREMENTS

While the level of total American commitment and involvement (both official and private) in the countries of the Transcaucasus and Central Asia is likely to increase gradually during the years ahead, no major new initiatives appear probable or called for. Pure military considerations are not likely to be a major factor in these relations. Five kinds of military needs can, however, be envisioned:

1. Air transport for humanitarian and emergency relief operations and, possibly, in support of peacekeeping arrangements

2. Provision of personnel for routine relations in the framework of the attaché system and also, possibly, for occasional special advisory and survey tasks.

3. Possible provision of personnel on at least a token basis for participation in peacekeeping missions

4. Provision of communication support for several possible kinds of missions

5. Provision of training slots in U.S. schools for Caucasian and Central Asian officers.

EXTREME BUT UNLIKELY CONTINGENCIES

From the viewpoint of the present, none of the contingencies listed below could be regarded as much more than remotely possible. They will, therefore, not be discussed in detail but merely be listed. Some would require urgent American diplomatic initiative; others would have serious military implications. All would require major adjustments in U.S. policies that extend well beyond considerations that would affect the Caucasus or Central Asia:

1. Election of Zhirinovsky or another neoauthoritarian or neo-imperialist as Russian president with a shift to highly aggressive policies toward Eastern Europe, Turkey, Iran, Afghanistan, Pakistan, and India

2. Spillover of one or more aspects of Caucasian strife into Turkey, necessitating a Turkish military response, and activating Turkey's entitlement to protection from NATO

3. Russian-Kazakh ethnic conflict in northern Kazakhstan

4. Spillover of Tajikistan's ethnic conflicts into northern Pakistan, western China, or other parts of the Himalayan region

5. Disintegration of China along lines comparable to what happened in the Soviet Union in 1991, with far-reaching implications for the countries of Central Asia as well as for Russia itself

6. Nuclear banditry of several possible kinds.

REFERENCES

Harris, George, *The Origins of Communism in Turkey*, Stanford, Calif.: Hoover Institution Press, 1967.

Henze, Paul, "The Great Game in Kashgaria," *Central Asian Survey*, Vol. 8, No. 2, 1989, pp. 61–95.

Hopkirk, Peter, *The Great Game: On Secret Service in High Asia*, London: John Murray, 1990.

Monterey Institute of International Studies, CIS Nuclear Database, Monterey, Calif., May 1, 1994.

LATIN AMERICA
Kevin M. O'Connell

From the Zimmerman Telegram to Pearl Harbor to the demise of the Soviet Union, American strategists had to be concerned that an extrahemispheric adversary might be able to secure a base within the Western Hemisphere to use as a platform to project a threat against the United States or its ability to reinforce allies along the Eurasian periphery. This strategic possibility was illustrated by events such as the intrusion of German submarines into the Caribbean during World War II and the Soviet deployment of nuclear-capable missiles to Cuba in 1962. With the disappearance of the Soviet Union, this type of fundamental challenge to our national safety and capacity to project power no longer exists: The Western Hemisphere faces no external conventional military threat.

Meanwhile, within Latin America itself, some traditional conflicts and arms competitions—such as Argentina-Brazil, Argentina-Chile, and among the Central American countries—are quiescent, but the border war between Peru and Ecuador, and occasional tensions within the region demonstrate the potential for nationalist senti- ments and misunderstandings to erupt into war. Latin America, in spite of those disturbances, remains an important force in world stability. The austere budget environment for Latin military organi- zations makes large-scale purchases of weapons and consequent arms races difficult, notwithstanding the availability of arms bargains from Russia and Eastern Europe. At the moment, only Chile has a military modernization effort of breadth and consequence under way; Mexico's military modernization to confront an uncertain challenge in Chiapas is in doubt, given the government's financial difficulties.

While, in other areas of the world, the proliferation of long-range air-craft; ballistic missiles; and nuclear, chemical, and biological weapons of mass destruction is posing more serious problems, the danger of these developments appears to be receding in Latin America. Argentina has abandoned the Condor II missile program and joined the Missile Technology Control Regime. Brazil has also indicated that it will adhere to the regime's export guidelines and will seek passage of tighter domestic export control regulations.

With regard to nuclear weapons, Argentina and Chile have deposited their ratification of the Tlatelolco Treaty—which provides for a Latin American nuclear-free zone—and Brazil has agreed to abide by the treaty. (Cuba has also signed the Tlatelolco Treaty.) Moreover, Brazil and Argentina have completed ratification of full-scope safeguard agreements between themselves and the International Atomic Energy Agency. In the 1993 Mendoza Declaration, Argentina, Brazil, and Chile pledged to ban chemical and biological weapons. Other Latin American states are expected to follow suit.

This propitious international state of affairs has been made possible by a convergence of opinion among Latin American elites. Latin America remains the largest democratic region in the world, rein-forced by national elections in Argentina, Brazil, Peru, and Uruguay (among others) and by participation in the broad hemispheric agenda agreed to at the Summit of the Americas in December 1994. Current consensus centers on the promotion of free markets, increased political and economic integration, and the promotion of democracy.

Economic policy within the region has centered on macroeconomic stabilization, free-market reform, and regional and international trade and economic integration within a democratic framework. The primacy of economics within the Latin democratic calculus was clearly demonstrated in Brazil, as Fernando Henrique Cardoso, the former finance minister and architect of Brazil's successful anti-inflation strategy, catapulted over a widely popular candidate in the presidential elections of October 1994. One of the key successes of

Summit of the Americas was general agreement on development of a hemispheric free-trade area.

Economic integration has progressed markedly within Latin America, increasing these nations' dependence on each other and on the United States. While the consequences of this have been largely positive—due to increased trade—it has also in the Mexican peso crisis demonstrated the vulnerability of national economies to external developments. The positive effects are clear. Intra–Latin American trade grew between 1988 and 1992 by 79 percent. Among the MERCOSUR states (Argentina, Brazil, Uruguay, and Paraguay), for example, trade rose by 58 percent in 1993. Andean Pact (Bolivia, Colombia, Ecuador, Peru, and Venezuela) trade expanded over the last three years by 20 percent annually.

Regarding developed-country markets, the Latin American countries were sending more than 50 percent of their exports to the United States and Canada by 1990—up from 40 percent in 1970. During the same period, exports to the European Community declined from 48 percent of the Latin American total to only 22 percent.

In terms of Latin American imports, the situation was the same. Latin America purchased 55 percent of its imports from the United States in 1990—an increase from 50 percent in 1970. And while U.S. sales went up, the European Community's (now the European Union's) share in Latin American markets declined from 30 percent in 1970 to 20 percent in 1990. Neither did Japan make new inroads during that period. Between 1987 and 1992, moreover, Latin America's imports from the United States grew faster than U.S. sales to any other region of the world.

The integration of markets among the hemispheric states extends to the realm of finance, notwithstanding the Mexican financial crisis. U.S. investment in Latin America has been increasing now at 2.5 times the U.S. worldwide rate. In 1992 alone, U.S. investment in Latin America went up 16.6 percent. By the end of 1992, cumulative U.S. direct investment in Latin America—ownership by U.S. companies of all or part of Latin American firms—had reached $88.9 billion.

The Mexican financial crisis, and its effect on Latin and other national economies, characterizes the negative aspects of this increased integration. Private investment in the region was accelerating because of increased trade and economic reform. The dramatic response to the floating of the peso and continued internal political and military concerns in late 1994 and early 1995 required a sustained response by the United States and international financial institutions, in addition to actions taken within Latin economies. Any other failure within a Latin country or another external shock (such as rising global interest rates, falling commodity prices, persistent Latin trade deficits, or another oil crisis) could have a "domino effect" reversing positive changes within the region.

While the Mexican financial crisis affected private investment in Mexico dramatically, and throughout the region generally, this seems to have abated with the stabilization of the peso and timely Mexican payments of its obligations under the bailout program. Private investment in Latin America seems to be returning to precrisis levels, albeit gradually. While private investment is again improving, any internal unrest, coup, or a major policy failure in Latin America or the emergence of investment opportunities elsewhere (such as Europe) could again reverse this current flow.

The cascading effect of external shocks and individual country failure could repeat the pattern of the collapse of the early 1980s, which followed an interlude in the 1970s of strong Latin American economic growth financed by heavy international borrowing. At that time, because of domestic political pressures, many of the Latin governments were operating on deficit budgets, which produced inflation and subsequent high wage claims. Hit by an external shock of rising oil prices, higher interest rates in the industrialized countries, a global recession and falling commodity prices, several Latin American countries suffered grave debt problems. Ensuing capital flight and weak international competitiveness led to difficulty in servicing the previously accumulated foreign debt.

The public implication of this was that real wages and consumption fell significantly, un- and underemployment increased, and government capacity to fund social problems was sharply reduced. In some countries, particularly in Central America, guerrilla warfare broke out, eventually generating an international crisis and diverting significant U.S. political attention and resources.

A more hopeful sign for Latin America, however, is that more and more of foreign private capital inflow consists of long-term engagements as opposed to governmental borrowing. Of the $54.6 billion of outside capital inflow in 1993, fully $37.5 billion constituted net long-term transfers—a sixfold increase over net long-term capital inflows as recently as 1989.

These inflows are sustained by confidence-producing economic reform programs, which, while providing gains for many, also create short- and medium-term problems for some significant groups—potentially sapping support for reform. The trick for Latin governments is to maintain the critical balance in favor of reform by providing sufficient transitional benefits to affected segments of the population.

Those groups requiring particular attention are medium and small manufacturers and merchants (stung by the loss of protection due to opening of markets and by the difficulties flowing from the rise of domestic interest rates because of governments' need to "sterilize" the inflationary impact of foreign capital inflows), rural small holders (also damaged by trade openings), and the urban poor (hurt by the absence of job creation by medium and small enterprises, by reduced government spending on social services, and by rising public utility and transportation rates). The availability of governments to provide benefits to affected groups is constrained by budget stringency and by the diversionary effect of corruption present in many Latin American countries.

The difficulties suffered by groups either affected by reform or fearing the future consequences of reform have a bearing on stability. Members of the Latin American military officer corps, increasingly drawn from the middle and lower classes, are now even more sensitive to the plight of those hurt by the short-term consequences of trade opening, stabilization, and reform.

Within the military organizations of a number of countries, themselves affected by budget reductions that constrain salaries, allowances, military housing, and other construction, some elements already view the reform process and democratic politicians with skepticism. This skepticism increases as memory of the militaries' recent governmental failures fades and corruption among the

democratic civilian politicians becomes more widespread and notorious. Latin American militaries may see themselves as the targets,
or even the victims, of reform, prompting an authoritarian reaction.

Coexisting within the Latin American militaries with "authoritarian
populist" elements identifying the poor and lower middle class are
conservative officers linked with elite interests. The priority for these
officers is the maintenance of order and military prerogatives.
"Populists" sponsored two coups in Venezuela in 1992; pro-elitist
elements were notable, on the other hand, in the Haitian coup of
1991.

While it would be wrong to consider Peruvian President Alberto
Fujimori's "self-coup" of 1992 as a pro-elitist move, the military supporting it certainly gave priority to public order and the counter-
guerrilla struggle over democracy. The Peruvian "self-coup," it
should be noted, had the effect of reversing political, while accelerating economic, reform. Fujimori's example has inspired other civilian
leaders to consider ways in which to combine traditional authoritarianism with democracy. Guatemalan President Serrano's aborted
1993 "self-coup" attempted to use the antipolitician inclination of
the military as a means to free an elected president from the fetters of
an elected congress. Finally, while Mexico's military leaders have
been quick to disavow any political pretenses—the chances of a coup
d'état are virtually nil—they remain concerned about their country's
stability and the abilities of its political leadership during a period of
dramatic change.[1]

Illicit narcotics—and their impact on government, military, and
police elements—is an increasingly serious problem. While some
Latin American militaries have kept out of the antinarcotics fight,
others have become involved in fighting, and still others in protecting, this traffic. U.S.-Mexican cooperation on counternarcotics
issues has improved in recent years, although there are hints of nar-
cocorruption within police elements and perhaps even some connection with the assassination of the Institutional Revolutionary
Party's (PRI's) presidential candidate, Luis Donaldo Colosio.

---

[1]Hemispheric cooperation on avoiding security and civil-military problems was
strengthened through a broad dialogue on these and other issues at the Defense
Ministerial of the Americas in Williamsburg, Va., in July 1995.

Colombia, in the wake of increased U.S. pressure, is trying to demonstrate—with some success—that it is willing and able to prosecute the antinarcotics war, especially against the Cali cartel. Beyond its importance as a criminal business, narcotics trafficking will become increasingly important as a source of political influence, through the financial support it can provide.

In general, Latin America is in the midst of an active period of political and economic reform. Overall political and economic trends hold great hope for enabling the countries of the region—individually and collectively—to begin to approach the levels of stability and prosperity that characterize North America and Western Europe; nevertheless, such internal pressures as increased authoritarianism and protectionism, or external economic or other pressures, could derail or completely reverse these favorable trends.

POTENTIAL HOTSPOTS

Cuba

Cuba is *sui generis* because of a complex and changing politico-economic dynamic, the health and resolve of Fidel Castro, and changes in U.S.-Cuban relations. While the United States, pursuant to the Cuban Democracy Act, maintains its long-standing embargo as leverage to promote change, it seeks to alleviate the hardships of the Cuban people by permitting humanitarian assistance, improved telecommunications services, and enhanced information flow. Cuba remains a top-priority area in which U.S. military interests are engaged and where direct military action is possible in the future, in the wake of a leadership or other dramatic change.

Internal tensions prompted by a continuing decline in the standard of living have led to a loosening of the economic system. Deprivation and scarcity in Cuba reached monumental proportions with the end of Soviet subsidies (about $6 billion annually), including a 75-percent decline in foreign trade and increased unemployment. But the removal of restrictions on the use of U.S. dollars, the opening of additional sectors, including sugar, to foreign investment, and the reopening of the *agromercados*—where peasants can sell individually produced crafts and foods—have improved the economic outlook ever so slightly. While the potential for internal strife remains, especially in the wake of planned large-scale layoffs, Latin American history during the 1980s suggests that severe economic difficulties are unlikely to provoke radical change, absent an organized and foreign-supported guerrilla group or a rebellion within the armed forces.

Given Castro's lifelong obsession with defying the United States, he is unlikely to institute political and economic reform to effect an accommodation with the United States. An agreement with the United States on immigration, under which the Cuban government would discourage illegal departures, accept criminal elements back from Guantanamo, and avoid harassment of those seeking to emigrate to the United States, however, is a case of selective accommodation at a time when the Clinton administration and the Congress are alternatively trying to loosen and tighten the U.S. economic embargo against Cuba.

Political change in Cuba is now likely to come only by a military uprising or in the wake of the death of the Castro brothers, although the continuing emergence of civil society could embolden reform elements within the regime. U.S. military interests would be directly affected by these or similarly dramatic changes, such as a civil war, a pro-Castroite guerrilla war, or general chaos. The United States and other nations of the Western Hemisphere may be compelled to intervene militarily to try to establish a democratic civilian administration. Significant humanitarian, police, and administrative assistance would be required in this case. Alternatively, should economic conditions again decline precipitously, or generate new levels of inequality in Cuba, the United States might again be faced with a massive refugee crisis. Both of these circumstances seem to have abated in Cuba, for now.

Haiti

Haiti's present situation is fragile, and its future uncertain, even in the wake of Jean Bertrand Aristede's return to power at the hand of a U.S.-dominated multinational force (MNF) in September 1994. Over a year later, the restored government faces numerous problems in distancing Haiti from its culture of violence and from general conditions of political and economic distress.

United Nations (UN) Security Council Resolution 940 authorized the use of "all possible means" for the restoration of democracy in Haiti. On 18 September 1994, a 22,000-man force consisting of troops from 30 nations—including almost 20,000 U.S. troops—arrived in Haiti under generally benign military conditions. While the MNF was successful in ousting the military government, restoring general law and order, and establishing the basis for improved police and military establishments, the March 1995 transition to a smaller and lightly armed UN force raises the possibility that Haiti's still fragile security environment could unravel quickly and dangerously.

Security will be essential to the conduct of local and national elections in Haiti, including the presidential elections planned for December 1995. While development of a new and more professional security force is well under way, criminal elements are still involved in activities that threaten personal security and economic activities. The UN force will have to continue to control this type of violence while monitoring, deterring, and preventing more sophisticated threats to the Aristede regime.

Haiti's government is slow in implementing the ambitious economic plans designed to raise Haiti out of the position of poorest country in the Western Hemisphere; plans to privatize state-owned enterprises, for example, have prompted considerable discontent at the hands of the elites and divisions within Aristede's cabinet. Haiti's ability to absorb foreign aid is limited, and the attraction of private foreign capital has been and will continue to be difficult. Popular economic expectations have not been met, even under the renewed Aristede regime.

For the United States, Haiti represents a successful removal of an authoritarian regime and the return to power of a democratically elected one, an important example for the rest of Latin America. Equally important, however, is the domestic policy issue prompted by a wave of Haitian refugees to the United States, as experienced in the past and as considered for the future. The arrival of the MNF in Haiti achieved the important objective of stemming the flow of these refugees from Haiti; given the early difficulties experienced by the UN force, and the lack of dramatic progress on the economic front, this exodus could easily resume in the near future.

Mexico

Mexico is clearly in the middle of a political and economic transition that could have major implications for the United States. The passage of the North American Free Trade Agreement (NAFTA) and subsequent U.S. support to stabilize the Mexican peso have not only inextricably linked political and economic developments within the two countries but have also had an impact on the U.S. world position, given a consequent depreciation of the dollar.

It is safe to say that the consequences of the collapse of the Mexican peso are not yet fully known. Insolvency in the Mexican banking sector, corporate financial problems, and dramatically higher unemployment (over 1 million people have lost their jobs) and inflation are among the most difficult problems that confront the Zedillo government; meanwhile, only the wealthiest Mexicans can acquire goods and services that they were accustomed to having in the past. Yet economic decline has not perpetuated widespread political unrest within a generally complacent population.

Meanwhile, Mexico's unique blend of democracy and one-party rule is increasingly threatened—by President Zedillo's showing in the August 1994 presidential elections, by opposition political victories in key Mexican states, and by hints of corruption by past and current officials. President Zedillo, to his credit, has recognized the importance of political transformation and has undertaken electoral and other reforms designed to move Mexico closer to a fully participatory democracy. He has also undertaken measures to attack corruption and reform the Mexican justice system.

Military circumstances have also changed. With the emergence of the Zapatista National Liberation Front (EZLN) insurgency and the discovery of oil in the Lacandon jungle, the southern state of Chiapas—largely populated by indigenous peoples—has increased in strategic importance. Talks between the military and the EZLN appear to be deadlocked, and while the EZLN's permanent cadre is believed to be small, it has successfully exploited local political and economic sensitivities and has gained substantial support among the Indian population. Moreover, the EZLN's leadership has turned what might have been a local insurgency into a transnational phenomenon, by making use of modern communication systems—such as the Internet—to create propaganda and to gain global sympathy and support. Strong networks of nongovernmental organizations have grown around EZLN-related issues, often with the effect of keeping the Mexican government and military on the defensive publicly, while adding to the pressures to reform the Mexican political system. Meanwhile, an ambitious Mexican military reorganization and modernization program may be on hold because of the country's financial situation.

The historical U.S. role as a world power is due, at least in part, to its enjoyment of stability on its southern border. It is unrealistic to expect this phenomenon to change in the future. Unless the Mexican government's political and economic reforms are successful in bringing economic growth (if not stability) and greater political plurality to Mexico, new violence and expanded immigration to the United States are likely to take place. Illegal immigration is already up dramatically in the wake of Mexico's economic woes. Popular disturbances could also occur in U.S.-Mexican border regions where opposition inroads are significant.

Mexico has traditionally rejected improved bilateral military relations, in part for historical reasons. Parallel to enhanced U.S.-Mexican cooperation on immigration and counternarcotics issues, the United States might become involved in the modernization and training of the Mexican armed forces.

Venezuela

During the last five years, Venezuela has experienced great political instability despite (or because of) efforts to undertake economic reform. Massive urban riots took place in 1989, and two military coup attempts took place in 1982, in part because of these reforms (including an increase in bus fares). President Carlos Andres Perez was impeached on corruption charges, and following a weak interim presidency, President Rafael Caldera took office with expressed reservations about the process of reform.

Venezuela's problems flow from public resentment over cutbacks in public expenditures and the subsidies previously financed by the oil boom of the 1970s and the institutional borrowing of the 1980s. As in several countries, the Venezuelan congress—seeking to preserve these benefits—blocked President Perez' efforts to reduce the fiscal deficit and the role of the state. Parallel with these economic problems, a breakdown of law and order, failure of the judicial system, and a generalized sense of corruption created considerable discontent.

President Caldera's administration, in response to a banking crisis, has felt compelled to suspend some constitutional guarantees, institute price and currency controls, and take *de facto* control of the banking system. Continued popular demonstrations, contentious charges by President Caldera about other branches of government, limited successes in an anticorruption program, and activities by "authoritarian populist" elements within the armed forces make the future of Venezuelan democracy uncertain, if not problematical. Economic policy is having little effect on stability or growth, according to the World Bank, signaling a continuing recession and inflation.

Venezuela is the second largest source of oil for the United States (15 percent of imports) and has the only surge capacity outside the Persian Gulf. Its viability as a democratic partner has important implications for U.S. energy security. Moreover, U.S. forces might play an important role in assisting Venezuela with peacekeeping missions outside Venezuela.

Panama

As the United States considers the strategic importance of the Panama Canal, its associated military presence in Panama, and its obligations under the 1977 Panama Canal Treaties to withdraw all forces from Panama by the end of this decade, it will do so in the context of a regime shadowed by its past connections to Manuel Noriega. The election last September of Ernesto Perez Balladares ("El Toro") signals the potential for renewed tensions with Panama at a critical juncture. President Perez Balladares has already allowed the systematic return to power of many former associates of the Noriega regime and earlier remnants of the Democratic Revolutionary Party; the government's decree of a day of national mourning upon the fifth anniversary of the U.S. invasion (Operation Just Cause) signaled a stronger and deeper message.[2]

The Panama Canal remains an important strategic asset for the United States, notwithstanding its size limitations. About 12 percent of U.S. seaborne trade uses the canal en route to and from Asia. In the event of Middle East and Far East contingencies, large amounts of U.S. military supplies would move through the canal. The transfer of large bodies of troops by ship from the Pacific to the Atlantic is roughly 15 days faster if the canal is available for use than if it is necessary to go around Cape Horn.

President Perez Balladares' approach thus far and his strong nationalist tendencies do not bode well for Panama's consideration of a U.S. presence beyond the year 2000, which the treaty provides for upon mutual agreement. The evolution of U.S. bases in Panama, other military issues, or separate issues, such as crime and counternarcotics, may also have an important impact on U.S. interests in the critical years up to 2000 and beyond.

Brazil

Brazil has moved smartly under President Fernando Henrique Cardoso to reform the tenth largest economy in the world, and by far the largest in Latin America. His election in October 1994 over the populist social democrat Luis Inacio "Lula" da Silva is widely attributed to measures introduced to stem rampant inflation and inject economic liberalization into a state-dominated economy. Looking beyond economic policy—where Cardoso is introducing reforms of the tax system and the social security system, curbing public expenditures, and liberalizing (but not immediately privatizing) the oil, health, banking, and telecommunications sectors—the new president has undertaken a virtual rewrite of Brazil's complex constitution, a major determinant in the government budget and in restrictions on foreign investment. A new Brazilian congress seems

---

[2]Recent developments, including a meeting between Presidents Clinton and Perez Balladares, may foreshadow serious discussions about a limited U.S. military presence and the future of the Panama Canal.

determined to overcome traditional fragmentation and work with the new president, in light of economic progress.

Brazil's recent successes do not necessarily foretell guaranteed progress, however; past successes against inflation and in economic reform have often been temporary. While Brazil, spurred by foreign capital inflows, experienced 4.5 percent growth in 1993, its growth between 1990 and 1993 averaged only 0.1 percent. This stagnation not only did nothing to help correct Brazil's extreme income inequality, but it also contributed to the plight of the 40 percent of Brazilians who live below the poverty line. Economic reform will have to extend to Brazil's impoverished if the country is to reach its true potential. However, events like the strike by workers at Petrobras, the state-run oil monopoly, demonstrate the threat of reform to powerful trade unions and one of the most difficult challenges of liberalization.

Brazil has played a positive role within Latin America, taking a lead role in the negotiations to halt the Peru-Ecuador war. (Brazil, Argentina, Chile, and the United States were involved as guarantors of the 1942 Rio Protocol). Yet the Brazilian military will remain a continuing concern for the United States. There is a widespread belief within the Brazilian officer corps that the U.S. desires to appropriate the Amazon region and encircle Brazil militarily. Despite the far-fetched nature of this belief, the Brazilian military works to reduce and constrain the U.S. military presence in neighboring countries. Brazil's military also does not wish to become involved in the counternarcotics struggle.

One way to counter these attitudes would be to develop greater cooperation with the Brazilian military, perhaps through joint work in extrahemispheric peacekeeping missions. Another problem posed by Brazil is its capacity to develop sophisticated weapons and to sell them to radical states (as it has done in the past to Iraq). Only the development of a broader consensus with Brazil on world politics will help alleviate this problem.

DRUG INTERDICTION

The Clinton administration has downgraded its predecessor's emphasis on drug interdiction, focusing instead on improving producer countries' counternarcotics institutions (e.g., enforcement and civil-justice organizations), funding sustainable development programs in key producing and transit countries, and enhancing worldwide law enforcement efforts to target narcotics "kingpins" and their organizations (through evidence sharing, controls on the flow of chemicals and cash, and extradition). Congressional initiatives to link counternarcotics progress to overall U.S. aid programs—as inserted into draft legislation for Colombia—and on the sharing of real-time intelligence on suspected drug trafficking flights (in abeyance of an administration legal finding) may offset this shifted emphasis.

Nevertheless, it is unclear what role the U.S. military will be called upon to play in Latin American counternarcotics operations. Legislative action may permit the military to resume an active role in helping Andean nations monitor and suppress air activity by narcotics traffickers. It is also possible that more vigorous military action may be required to help those nations fight the drug trade.

CONCLUSION

While there are many promising signs in present-day Latin America—the growing consensus on political and economic reform, the opening of markets, and the apparent return of a steady flow of private capital, following the Mexican financial crisis—a number of sources of instability remain. Economic progress in the past (especially in the post–World War II period and in the 1960s) has been followed by a return to authoritarianism, populism, and protectionism. The Peru-Ecuador war signaled that, as in many other parts of the world, nationalism can resurface as a dominant force in crisis and conflict.

Mexico plays a unique role for the United States, in light of NAFTA and the U.S. role in assisting the Mexican government through a difficult period, the political and economic consequences of which are still very uncertain. Because of increased U.S. trade and financial engagement, other situations could arise in the region that would affect significant U.S. interests. Immigration will also play an important role. Our security policy toward Latin America will naturally evolve from the Cold War preoccupation with containing military threats back to the more expansive 19th century notion of protecting

American commercial and financial opportunities and the newer problem of preventing attacks on the U.S. social fabric from illicit narcotics and those who provide them. This new, or revived, policy will have to rely on well-orchestrated political, economic, and military instruments to shape an environment conducive to the protection of U.S. public and private interests. As in Asia, an active U.S. security role and the presence of U.S. forces can help maintain a sense of stability and prevent the reemergence of struggles for power among Latin American states.

Chapter Twelve

SOUTH ASIA

Ashley J. Tellis

The South Asian region is defined as the Indian subcontinent, that is, the geographic landmass bounded principally by the Hindukush, Karakoram, and Great Himalayan ranges in the north and the Indian Ocean in the south.[1] By all indices of national power, such as physical size, population, economic strength, and military capability, the most prominent state in the region is India. It is followed, in terms of such indices, by Pakistan, Bangladesh, Sri Lanka, Nepal, and Bhutan.[2] Adjoining states with political relations with the region— Afghanistan, Myanmar, and China—lie outside the geophysical boundaries of the subcontinent, although China, with its large concentration of economic and military capabilities, possesses a "virtual" presence in South Asian geopolitics that is hard to ignore.

This chapter focuses on appraising the geopolitical environment in South Asia, emphasizing the grand strategies pursued by the states of the region. It will concentrate mainly on India and Pakistan, while referring as necessary to the security postures of other South Asian states and to the influence of other key external actors. The chapter is divided into three sections: The introductory first section describes the security strategies of the South Asian states since the British divested political control of the subcontinent in 1947. The second section describes how these strategies have evolved in the

---

[1]This chapter is a shorter version of a forthcoming RAND study on grand strategy and aerospace power in South Asia.

[2]For a comparative survey, see Robinson (1989).

post–Cold War period and how the regional states are attempting to come to terms with a new and radically altered world order. The third section summarizes the main trends and examines the implications of political developments in South Asia for the United States, including an assessment of the potential role of U.S. air power with respect to the region.

THE GRAND STRATEGIES OF THE SOUTH ASIAN STATES DURING THE COLD WAR

The South Asian region is defined by a unique civilizational entity that contains a multiplicity of languages, religions, ethnicities, and cultures, but has never experienced political unity of the kind seen in European nation-states. Throughout its history, subcontinental empires have competed with regional kingdoms; the British *Raj* was the most recent South Asian experience of a unifying subcontinental empire, and its dissolution in 1947 gave rise to two successor states, India and Pakistan.

Each of these new states represented a novel endeavor in South Asian politics. India sought to transform a multicultural empire into a unified secular state governed by liberal principles, while Pakistan attempted to consolidate linguistically and ethnically disparate groups into a single state based on a common religion, Islam. These states became competitive from the beginning because each sprang from a deeply held premise that challenged the other's legitimacy: Pakistan, born of the insecurity of some South Asian Muslims, challenged India's claim that its secularism was genuine enough to allow different religious, linguistic, and cultural groups to survive and flourish within it. On the other hand, if India were successful in maintaining a free political system that allowed its various groups to live together peacefully and prosperously, it undercut the reason for which Pakistan was established in the first place.

From India's point of view, the creation of Pakistan affected its strategic prospects in multiple ways. To begin with, it upset the natural geographical unity of the subcontinent by creating a new military threat, now arising from within, in addition to those dangers traditionally seen as emerging from without. Further, it complicated Indian efforts at unifying its diverse regional, linguistic, and cultural subgroups by serving as a source of both material assistance and ideational inspiration for various separatist claims. And, finally, it forced India to allocate economic and military resources in a struggle for political hegemony *within* the South Asian region when it could well have otherwise been allocating such resources to pursue a larger extraregional and, perhaps, even a global role. For these reasons, Pakistan represented the principal impediment to India's core grand strategic objective: *thriving as a great power, with all the security accruing from the possession of that status.*[3]

It is important to understand that India, being heir to an ancient civilization, possessing a large population and an extensive land-mass, and having great economic, technological, and military poten-tial, conceived security essentially to mean *existence as a great power.* True security could derive only from an unchallenged recognition of its standing as an important state about to actualize its vast potential after several centuries of division and subjugation. Thus, political *in*equality was the only appropriate principle within the South Asian region. And regional harmony could exist only when the smaller states, recognizing the "natural" imbalance of power within the sub-continent, chose not to contest Indian claims to hegemony.[4]

India adopted a multipronged strategy to sustain its relative power position. First, it embarked on an autarkic economic strategy designed to create the industrial and technological capabilities required to sustain both defense and development goals with mini-mum external assistance, paying particular attention to the high technology, atomic energy, and space sectors, which were salient for the purposes of power politics.

Second, India used its relatively larger resource base to deploy large armed forces capable of defending contested territorial claims as well as meting out some measure of punishment on neighboring states

---

[3]Good discussions of the enduring principles beneath Indian strategic and foreign policies can be found in Bandhyopadyaya (1980) and Tharoor (1982).

[4]The term "hegemony" throughout this chapter is used purely in a descriptive rather than a normative sense. It is meant to indicate a quality of political dominance that derives from the ability of one state to overwhelm its political competitors decisively—and effortlessly—through military means.

that might seek to alter the status quo through force.[5] This entailed primarily an emphasis on the army: A large army exploited India's comparative advantage in manpower while also being useful for internal security tasks and "state-building." India also maintained a relatively large air force, importing weaponry from Western Europe and the Soviet Union. The navy, which contributed little to the political outcomes within the subcontinent, was traditionally neglected (Tellis, 1987, pp. 185–219).

Third, India pursued a political strategy of nonalignment, which was intended to maintain its freedom of action with respect to the great powers. But because the international structure was bipolar during the Cold War, India eventually developed a close relationship of convenience with the Soviet Union as a counterbalance to Pakistan's episodic alignment with the United States and its close and growing relationship with China. This enabled India to secure large quantities of relatively sophisticated military hardware at favorable terms while simultaneously providing diplomatic and political cover against U.S. and Chinese pressures. It also served to ward off potential Soviet overtures toward Pakistan.

Finally, India avoided creating any regional security forum in which the smaller states might gang up against India,[6] even as it sought to dissuade extraregional powers from getting involved in the security competition within the subcontinent. The goal was to isolate the subcontinent politically; then India's relative power superiority, remaining undiluted, could be brought to bear within the region whenever necessary and outside it whenever possible.

From Pakistan's point of view, the hegemony that guaranteed India permanent security appeared menacing. In large part, this was because Pakistan believed India had never come to terms with Pakistan's existence or its self-image as the guardian of the region's Muslims. India was consequently perceived as being willfully determined to "undo" the partition of the subcontinent and, by

---

[5]A succinct summary of Indian conceptions of its armed forces and their utility can be found in Cohen and Park (1978), pp. 13–24.

[6]For this reason, India strenuously (and successfully) objected to the South Asian Association for Regional Cooperation (SAARC) becoming a forum for discussion of political disputes. The origins and structure of SAARC are described in Mendis (1991).

implication, to end Pakistan's independence. The core objective of Pakistan's grand strategy was, therefore, simply to survive in the face of its larger, more capable, and ostensibly revisionist competitor, India, with as much of its autonomy intact as possible.

Such an objective was not infeasible, given Pakistan's own capabilities, which were fairly substantial. To begin with, it deployed a relatively smaller but still capable military force, which, although unable to defeat India in a long, open-ended war of attrition, could nevertheless more than hold its own in any conflict of short duration, especially one whose termination was enforced by extraregional intervention.[7] For this purpose, Pakistan developed a capable army, with strong infantry and armor components deployed as close as possible to the border with India, and a small but highly trained and well-motivated air force, equipped with the best aircraft available to Pakistan for both air defense and ground-attack duties. The navy was strategically irrelevant in this context and was consequently neglected.

Such a force structure placed a premium on seizing the initiative. Pakistan's traditional military strategy required the offensive use of its army and air forces under conditions of strategic warning. If conflict was imminent, Pakistan would strike preemptively to secure small portions of Indian territory. These gains would then be used either to deflect the weight of Indian counteraction toward the recovery of its own territory or as a bargaining chip to secure a favorable negotiated outcome in the postconflict phase.[8]

Politically, Pakistan, in sharp contrast to India, sought extraregional allies to secure arms, war materials, and diplomatic support and looked to those allies as political guarantors who could intervene on Pakistan's behalf *in extremis*. The Cold War provided a hospitable environment for such a strategy. Pakistan initially aligned itself with the United States through the Southeast Asian Treaty Organization and the Central Treaty Organization. The failure of the United States to support Pakistan in the 1965 Indo-Pakistani war demonstrated the limits of such a strategy. The United States was interested in

---

[7]The logic of Pakistani *military* strategy has been analyzed in some detail in Tellis (1986), pp. 264–268.

[8]Tellis (1986), pp. 264–268.

Pakistan as a client insofar as it advanced the larger objective of containing the Soviet Union, whereas Pakistan sought U.S. assistance primarily in support of its security problems with India. Consequently, China—as a result of its own security competition with India—and the Islamic states of Southwest Asia—for ideological reasons relating to Muslim solidarity—appeared as Pakistan's newest allies. The Soviet invasion of Afghanistan provided another opportunity for renewing the U.S.-Pakistan relationship, but it foundered on the evidence of Pakistan's nuclear program in the aftermath of the Soviet withdrawal from Afghanistan.

By the end of the Cold War, the strategic competition in the Indian subcontinent was not decisively resolved in favor of one or the other state. India managed to split Pakistan in half during the 1971 conflict, producing the new state of Bangladesh. But it did not acquire the hegemony that it believes would provide real and lasting security. Despite almost 50 years of economic, technical, and military investments, India arguably still does not have decisive war-winning military capabilities, even when measured by raw numbers. When the weight of its power is divided by the extent of its defensive perimeter and the variegated nature of the demands made on its armed forces, India's numerical superiority appears evanescent. And when this power is evaluated in terms of the war-fighting outputs attainable on the battlefield, the "hegemonic" capabilities sometimes attributed to India all but vanish.

To the degree that such "hegemonic" capabilities exist at all, they are manifested only with respect to the smaller states, such as Bangladesh, Sri Lanka, Nepal, and Bhutan. Bangladesh occasionally engages in rhetorical contests with India over ongoing disputes related to Indian support of the Chakma insurgency, control over water levels at the Farraka Barrage, and ownership of the riverine islands in the Ganges delta, but its immense internal problems prevent it from engaging in serious strategic competition with India. Sri Lanka, through Indian involvement in the Tamil insurgency of the past decade, has recognized that its international preferences can be sustained only to the degree that they do not seriously undercut Indian regional interests. Nepal was painfully reminded of its vulnerability when it tried to assert its autonomy in the late 1980s by purchasing some antiaircraft artillery and other small arms from China. In response, India imposed a virtual blockade of Nepal, which ultimately promised to be more sensitive to Indian security concerns vis-à-vis China. Bhutan's dependence on India is codified by treaty relationship; more importantly, it relies on India for economic and physical survival and cannot imagine any other political alternative.

Thus, at the end of the Cold War, India has gained primacy in South Asia, but not hegemony; Pakistan has survived as a political entity, but still does not feel safe. As a result, it repeatedly falls prey to the temptation of supporting various separatist movements within India. Supporting these separatist movements is intended to keep India "off-balance"; as a strategy, it is based on the premise that diverting Indian attentions toward domestic security challenges purchases additional security for Pakistan while it struggles to develop more permanent antidotes to its fears, such as nuclear weaponry. India, in turn, has traditionally responded by supporting various separatist movements within Pakistan (though not on comparable scale), even as it continues to develop its own nuclear weapon capability aimed at deterring both Pakistan and China. The continuation of low-level proxy warfare in Kashmir and Sind (and, to a diminishing extent, in the Punjab) and the incipient nuclearization on both sides reflect the precarious equilibrium presently existing in the subcontinent.

REORIENTING GRAND STRATEGY AFTER THE COLD WAR

There is little doubt that Indo-Pakistani competition throughout the Cold War took a considerable toll on both states, and such competition has by no means ended. But the changing international structure and the alterations in the domestic environment in both countries have forced both states to recognize that their traditional strategies have reached the limits of their success. Both India and Pakistan have been forced to confront four new structural realities—two external and two domestic—as they reorient their grand strategy.

The first, and most obvious, new structural reality is the disappearance of U.S.-Soviet competition, which both India and Pakistan exploited. Neither state can now count on the automatic assistance of one or the other superpower as they did in the years of the Cold War. Because of the demise of the Soviet Union, the United States has fewer incentives to support Pakistan militarily in its security competition with India: Pakistan's most important supplier of critical military technologies no longer exists as such.[9]

By the same token, India has lost its most important patron. Soviet diplomatic, political, and material support—which was crucial both for maintaining Indian military capabilities vis-à-vis China and Pakistan and for supporting Indian efforts at increasing its military influence in the wider Indian Ocean region—is now unavailable. Unlike the rupee-barter transactions of the past, India's new arrangements with military suppliers in the CIS now require scarce hard currency; even worse, they only provide technologies that are becoming rapidly obsolescent. But India and Pakistan face a new reality: The single remaining superpower, the United States, has all the military, technological, and economic resources they covet but lacks the incentive to make these freely available to either or both states.

The second external reality is the emergence of China as a potential great power. So far, the most significant evidence of this has been the explosive growth of its economy. During the last 15 years or so, China's economy has grown annually at an estimated rate of approximately 9 percent and, by one calculation, has already achieved a GNP equal to one-quarter of that of the United States.[10] Such growth not only dwarfs India's economic performance, but it is unsettling to New Delhi because it appears to be accompanied by a Chinese effort at modernizing its military capabilities. This modernization involves developing some power-projection capabilities, but more importantly from an Indian perspective, it also involves reorganizing the requisite land and air ("fist") forces necessary to prosecute limited, high-intensity wars. Since this modernization is aimed at defending contested territorial claims along China's littoral and border regions, it portends serious Indo-Chinese military-strategic competition down the line—the present "normalization" in ties notwithstanding. It places increased pressure on India to revitalize its economy to avoid becoming disadvantaged in an age when external superpower assistance is no longer automatically available.

---

[9]The importance of the United States in traditional Pakistani security strategies is analyzed in Cohen (1985).

[10]"China: The Titan Stirs" (1992).

The fact that India might succeed in this revitalization effort, in turn, unnerves Pakistan because it implies that more economic resources will be available for Indian military purposes at a time when Pakistan finds itself desperately short of reliable allies. The recent Indo-Chinese efforts to improve relations only increase the traditional Pakistani fear of isolation even further (Nuri, 1993, p. 94). While the durability of the Chinese tie is not yet in question, there is a clear recognition that China will not be a source of high-quality military or civilian technology well into the distant future. The ability of the moderate Arab states and the new Central Asian republics to help in this regard is similarly limited; the major second-tier powers, such as Europe and Japan, are seen as being sufficiently dependent on the United States that no substantive assistance is possible without U.S. acquiescence.

At the same time, both India and Pakistan have to deal with increasing internal pressures that have taken on qualitatively new dimensions. The third important reality that India and Pakistan face in this regard is the failure of their traditional economic strategies. India followed a strategy of relying on a centrally directed economy in the hope that it would advance both power-politics and development goals. Neither objective has been satisfactorily obtained. In India, the state sector has been generally characterized by poor management, pervasive inefficiency, and low rates of return. This underperformance has been exacerbated by operating in a milieu bereft of internal and foreign competition.

As a result, the Indian economy has neither been able to generate the requisite surplus necessary for steady and high levels of growth nor has it been able to encourage the kind of innovation required to sustain the development of new technology continually. This has been true for both civilian and defense industrial sectors. India has not succeeded in producing advanced goods competitively or in solving the fundamental problems of development, especially with respect to its marginalized population. Neither has it managed to develop autonomously the kinds of sophisticated technologies required to make it militarily self-sufficient. Indian defense industrial capabilities are extensive but its high-technology sectors are for the most part either licensed producers of foreign arms or centers for reverse-engineering and modifying imported technologies at the subsystem level.

Pakistan faces even worse problems in this regard. A large semifeudal agrarian economy coexists with a much smaller state sector that is as inefficient as its Indian counterpart. As a result, low savings and investment rates are the norm—a problem only complicated by high budget deficits, a narrow tax base, high defense expenditures, and low investments in infrastructure, health, and communications. Like India, Pakistan is able to produce a variety of small arms indigenously, but it is even more severely dependent on external technology for big ticket items, such as tanks, aircraft, and ships. These weapons are invariably purchased off the shelf, with only maintenance undertaken domestically. Similarly, all guided munitions are imported, making Pakistan highly vulnerable to producer cutoffs in wartime. The bottom line, therefore, is that Pakistan has few or no autonomous defense industrial capabilities of consequence.

The dramatic economic liberalization now under way in India, and to a lesser extent in Pakistan, is thus significant because it represents a response to crisis. Both states have begun to decontrol the internal economy and to increase linkages with the global economy in the hope of securing both capital and high technology for civilian and military purposes.

The failure of traditional economic strategies has coincided with the recognition that both states have to pay increased attention to domestic regime maintenance if they are to continue to survive. This is the fourth new structural reality, which takes different forms in India and Pakistan. In India, the question of regime maintenance increasingly consists of satisfying increased societal demands for economic justice and political participation. These demands are in some cases simply an outgrowth of modernization. In other cases, they arise from the failure to apportion growth equitably in a situation in which the mediating structures within the democratic political system have atrophied. Irrespective of the causes of social unrest, its net effect has been to increase the difficulty of maintaining the current political regime. Indian policymakers have responded to this challenge by focusing on renewing the domestic economic base even as they struggle to pacify the numerous secessionist movements.

In Pakistan, the question of regime maintenance is intimately linked with the issue of transitioning to a permanently democratic political structure. This requires attenuating the power of the established military and civilian bureaucracies that control the armed forces and significant portions of the economy at large. Reducing bureaucratic power, in turn, requires, among other things, dismantling the network of economic controls and administered prices, restructuring the tax base, and promoting foreign direct and portfolio investments. Such an economic reorientation is necessary if Pakistani *society* is to be revitalized in the face of a powerful *state*. Efforts have begun, albeit in fits and starts, but the end result is still uncertain, given that the most prominent civilian leaders are still divided by partisan politics rather than united against entrenched military-bureaucratic interests.

Given this set of new circumstances, it is no surprise that both India and Pakistan have been forced to shift their traditional grand strategies. The old objectives have not been altered: India still desires hegemony as a means to security; Pakistan still desires survival with independence. What has changed, however, are the circumstances and, consequently, the means by which these objectives are pursued.

India has realized that the demise of the Soviet Union and the rise of China imply that any future claims to hegemony will confront potential Chinese opposition. China is the only proximate actor of consequence with the ability to erode Indian claims to preeminence both within South Asia and the Asian continent at large. For the moment, such rivalry is muted for tactical reasons. Both China and India need breathing room to complete the domestic economic transformations currently under way. The inexorability of long-run competition with China (together with the continuing rivalry with Pakistan), however, serves to remind New Delhi that Indian claims to regional hegemony cannot be sustained without acquiescence of the United States. Indian calculations here are straightforward. Both Chinese and Pakistani opposition to Indian hegemony can be contained if American support for India is forthcoming. But, should the United States choose to contain India actively—with or without the collaboration of others—all Indian aspirations to hegemony will be stillborn. The gap in power capabilities between India—a rising power—and the United States is simply too great for any other outcome to be feasible.[11]

Recognizing just this fact, New Delhi's principal objective in the post–Cold War period has been to co-opt the United States into accepting an Indian managerial role in South Asia. Pursuing this objective has been eased by the American desire for broader cooperation. At present, U.S. objectives consist mainly of "testing the waters": India is a populous state with fairly significant military capabilities. It lies astride the sea lanes of interest to the United States, and it abuts China, a potential threat to larger American interests over the long term. From India's point of view, however, the diplomatic challenge consists of essentially leveraging this limited U.S. interest in India into a broader strategic acceptance of India's "stabilizing" role in the region at large. Toward that end, India has sought to position itself as a buffer against rising Islamic "fundamentalism"; a constraint on Chinese hegemonic aspirations; a source of support toward those of America's Southeast Asian allies potentially threatened by China; and a satiated "status quo" power that seeks to defuse the global problems of nuclear addiction and proliferation, terrorism, and diffusion of weapons of mass destruction. In short, by focusing on the issues on which Indian and U.S. interests might converge, India has sought to depict itself as a state whose friendship would advance larger U.S. strategic interests along the wider southern Asian rim.

Irrespective of how justified this position ultimately may be, the fact remains that India presently needs U.S. support along at least two dimensions. The first dimension is economic: India could use increased U.S. investment in its capital-starved economy. Equally important, however, is the *form* in which this new investment takes place. In a sharp departure from the strategy of the past, India now seeks new U.S. investment principally in the form of transactions carried out by the private sector in both countries. This new emphasis on the private sector is critical, because it reflects a conscious, but understated, strategy of advancing Indian grand strategic objectives.

---

[11]An instructive analysis of the gap in capabilities between India and various Western states, including the United States, can be found in Panchamukhi (1991), pp. 212 ff. For a brilliant theoretical analysis of the relationship between a hegemonic state and rising challengers, see Nayar (1979).

It allows American private enterprise to underwrite partially the costs of Indian economic growth through capital and technology infusions in critical areas, such as power generation, communication and information networks, and industrial high technology; it provides the Indian state with access to sophisticated technologies (some of which have clear military spin-offs) without risking the political problems that would arise if it tried to acquire such technologies directly; and, finally, it creates a domestic pressure group within the United States with a vested interest in continued economic ties with India irrespective of developments in other strategic issue areas. Thus, increased American private investment in India contributes to Indian economic growth and helps produce the investible surplus that the Indian state can use for its own purposes—including power politics.

In addition, U.S. support is sought directly in the military arena as well. The Indian armed forces are large and diversified, but their equipment has severe limitations; their war fighting doctrine is antiquated; and their domestic research, development, and innovation base is hardly state of the art. Even if the Soviet Union had *not* disappeared as a steady supplier, the Indian armed services would still be equipped with relatively inferior weaponry. The truth of the matter is that the Indian armed services have barely perfected the art of fighting wars of attrition at a time when the best Western combat armies are on the verge of leaving even maneuver warfare behind in favor of long-range precision interdiction based on information dominance.[12] The U.S. performance in the Gulf War came as a shock to the Indian military.[13] The technological inferiority of Soviet

---

[12]This development, sometimes referred to as the "revolution in military affairs" or the "military-technical revolution," is analyzed in relationship to previous revolutions in Arquilla (1994).

[13]Only a week before the ground war, for example, one of India's foremost strategists, Gen. K. Sundarji, remarked that "from the Iraqi point of view, looking at the quality of the ground troops of the alliance should give them heart. For barring the Pakistani and British troops with their regimental traditions, and the U.S. Marine Corps with their traditional élan, I do not think the other infantry and tank soldiers will be very much better than the Iraqis." See Sundarji (1991). The events of the following week would demonstrate that the morale of the fighting soldier, no matter how high to begin with, could not be maintained in the absence of operational flexibility, superior technology, or effective doctrine. For a synthesis of early Indian readings of the "lessons" of the Gulf War, see Garrity (1993).

weaponry, combined with the operational flexibility of Western combat arms, suggested that the attrition strategies followed in the subcontinent would be fatal against any adversary armed with modern technology, doctrine, and tactics. Consequently, it is no surprise that the Indo-U.S. *pas de deux* in the defense arena has been substantially encouraged by the Indian armed services.

While civilian policymakers are still reticent about being too visibly associated with the United States for reasons related as much to domestic politics as to innate suspicions of long-range American intentions, the Indian armed services have clearly seen the benefits of increased Indo-U.S. defense cooperation. The Indian navy and air force, the two services that use technology most intensively, would benefit most from a wide-ranging dialogue on technology transfer, force exercises, training exchanges, and military-to-military consultations. Such efforts have been promoted through the joint steering committee formed between the armed services on both sides. And this dialogue—though often lengthy, hesitant, and arduous—has now resulted in the first transfer, since the 1960s, of American lethal weapon technologies to India.[14] The bottom line, therefore, is that all three Indian armed services view the United States as a source of leading-edge military technology and innovative force-employment doctrine.

The reorientation in Indian grand strategy, therefore, consists of engaging the United States in a wide-ranging dialogue across the board, even as India tries to salvage relations with Russia, and attempts to foster broader contacts with second-tier states in the international system, such as Germany, Japan, Israel, and Italy.[15]

---

[14]It is reported that India will purchase 315 Paveway laser guidance kits for its British-designed 2,000-lb bombs and is considering a purchase of GBU-10 laser-guided bombs. These purchases, as well as the nature of U.S. participation in major Indian weapon development programs, such as the Light Combat Aircraft and MiG-21 upgrade programs, are discussed in Fulghum (1994).

[15]The visit of the Indian Prime Minister, P. V. Narasimha Rao, to Russia was a classic example of the Indian efforts at salvaging Indo-Russian relations. Despite a variety of agreements signed on that visit, including one on defense-related debt and another on coproducing MiG spares for export, it was evident that, as one report put it,

[T]his [visit] does not mean a revertal [sic] to the emotionalism of the special relationship of an earlier era that carried the stigma of an ideological alliance with a communist Moscow. Instead the current visit marked the development

The objective of this dialogue is to encourage the United States to commit as much as it can by way of investment, technology, and arms without having to give up on India's traditional policy of avoiding tight alignments. India seeks to be perceived as secular, moderate, and reasonable—a posture that presumably distinguishes itself from its competitor, Pakistan. Consequently, it has encouraged the United States both to permanently cut off arms supplies to Pakistan and to restrain Pakistan's incipient nuclear weapon program through a vigorous application of the Pressler Amendment.

Obviously, these are only near-term objectives. They make most sense in the context of a larger goal, which is to secure U.S. cooperation in ensuring the trouble-free attainment of permanent regional hegemony. Such hegemony, India believes, will come about sooner or later, as its economic revitalization gathers steam. The only question, therefore, from an Indian perspective, is whether the United States will facilitate that ascent to hegemony by recognizing India as a "regional manager." That would prevent local competitors from invoking extraregional protection and would thereby promote the traditional objective of enforcing the geopolitical isolation of the subcontinent. Even in its most recent shift in strategy, India continues to pursue this objective—only this time through engagement with the United States rather than by opposition to it, as was the case throughout the Cold War.

The Indian shift from opposing the United States to engaging it has unnerved Pakistan for obvious reasons. A potential Indo-U.S. alignment, especially in a unipolar environment, implies the isolation of Pakistan at a time when no other great power exists to provide compensatory cover. It coincides with a dispute with the United States that revolves around the Pakistani nuclear program. This program was pursued full steam throughout the 1980s; it was overlooked, however, by the United States so long as the Soviets remained in Afghanistan. Once the Soviet withdrawal was completed, Pakistan's

---

of hard-boiled sense of realism geared to the realization of the mutual benefit. (Basu, 1994, p. 10.)

That Russian interest in India is now clearly secondary to its broader interests vis-à-vis the West was demonstrated abundantly when, under U.S. pressure, Russia reneged on a promised sale of cryogenic engines to India. The implications of this event are analyzed in Alam (1994).

relevance to the United States became questionable. Consequently, the importance assigned to containing the Pakistani nuclear program grew in significance, and the United States responded to Pakistan's refusal to abandon that program by cutting off all conventional military assistance, including critical air and naval equipment, such as the 28 F-16s and 3 P-3C Orions.[16]

Pakistan's choices in these circumstances are not enviable. If it gives up the nuclear program, it will secure some critical military technologies in the short term but, without permanent security guarantees from the United States, will once again face the problem of securing reliable and capable allies. Given the recent Indo-U.S. rapprochement, Pakistan is uncertain as to whether the United States would be willing to underwrite its survival over the long term (Taqui, 1991, p. 6). The record of the past is seen as disheartening. The advantages of U.S. capability have been repeatedly negated by its political unreliability. China, on the other hand, presents the opposite problem. The Sino-Pakistani alignment will continue to endure, cemented by fears of India, if by nothing else. But China, although reliable, is incapable of aiding Pakistan in terms of either sophisticated technology or actual armed assistance and will continue to remain so for a long while. Giving up the nuclear program in these circumstances, therefore, appears problematical.

If, on the other hand, Pakistan retains its nuclear program, it is disadvantaged in other ways. It will continue to be denied important categories of American military equipment; thus, its modest indigenous nuclear capability will be maintained only at the price of serious conventional weakness. Such a situation is not reassuring to Pakistan for two reasons. First, the Indian nuclear program is much larger and much more diversified than its own, and the Indian nuclear arsenal could eventually be between three to five times larger

---

[16]For a brief survey of these developments, see Kemp (1993). With the recent passage of the Brown Amendment, the United States has—much to India's discomfiture—secured in effect a "one time" waiver of the Pressler restrictions affecting weapon sales to Pakistan. Under this amendment, the United States will transfer three P-3C Orions, limited quantities of Sidewinder and Harpoon missiles, mortar-locating radars, and towed artillery pieces, as well as some miscellaneous equipment for Pakistani aircraft and helicopters. The 28 F-16s, however, will *not* be transferred, although the United States is committed to returning Pakistani funds through third-party sales of these aircraft.

than its own nuclear force. In a crisis, such asymmetry in capability could lead to blackmail, especially if India chooses to exploit some version of escalation dominance flowing from its nuclear superiority. Second, Pakistan's conventional weakness means that the decision to go nuclear would come earlier rather than later should any conflict erupt. Without a robust conventional defense, Pakistani security would essentially become hostage to its ability and willingness to engage in nuclear *risk-taking* against a larger, more capable, and better diversified Indian arsenal. Retaining a nuclear weapon program, therefore, may not provide guaranteed security if in the end it produces only a small arsenal of doubtful survivability maintained in the face of severe conventional weakness.

Given that neither option is attractive, Pakistan's strategy presently consists of straddling the fence. It still seeks the best of both worlds—continued American assistance in maintaining robust conventional capabilities while pursuing its nuclear option—even though this is impossible to obtain. Consequently, Pakistani grand strategy is currently pursuing a multifaceted effort at "option enhancement." To begin with, it is attempting to restructure its economy in the hope that improvements in economic performance will provide the state with the resources needed to buy the best conventional arms available on the open market. If the restructuring strategy succeeds in this way, Pakistan theoretically could continue to pursue its nuclear program uninhibited by external pressures, while using its domestically garnered resources to buy advanced British, French, and Russian conventional weaponry through commercial sales. This strategy could be negated, however, if the United States chooses to engage in compensating arms sales to India, or if the Europeans and Russians adopt a policy of export restrictions or reductions in financial credits.

To avoid just these outcomes, Pakistan recognizes the need to engage in continued dialogue with the United States, even if the latter persists in its embargo on actual weapon transfers. The objective here is to remind the United States that Pakistan's permanent security interests consist not of regional hegemony but simply of ensuring its survival and autonomy vis-à-vis its larger neighbor, India, and that its close relations with the Islamic states of Southwest and Central Asia make it, rather than India, the logical partner for U.S. initiatives in these critically important, but potentially unstable, regions of the world. These twin arguments are intended to reinforce the claim that Indian friendship toward the United States is transient and instrumental: Being designed to enhance Indian capabilities, it will accelerate the movement toward multipolarity and advance the deterioration in relative American power. Pakistani policymakers would argue that, in contrast, a U.S.-Pakistani tie involves no such liabilities. It provides for the security of a small, moderate, Islamic state; it can serve as a check on larger Indian regional ambitions, especially in the northern Indian Ocean and in Southeast Asia; and it gives the United States increased access and influence among the many smaller regional states that would value the United States more than India does simply for reasons of strategic necessity.

Regardless of whether these arguments are ultimately true, the result is that the United States is in the happy position of being wooed by both South Asian adversaries, each using a set of arguments that mirror the other's. Aware of the fact that which side ultimately "wins" will be determined by the United States, Pakistan has sought additional forms of insurance. This insurance largely consists of continuing to develop its long-standing ties with China, complemented by a more intensive dialogue with Iran, Saudi Arabia, the various Central Asian republics, and, recently, Russia. None of these states, either singly or together, is seen as an alternative to cooperation with the United States at present. Yet, they represent the object of Pakistani efforts at developing a "fallback" position; to the degree that these relationships can be successfully cemented, they hold the promise of enhancing Pakistan's own value to the United States. Unfortunately for Pakistan, none of the constituent elements of its fallback strategy have been entirely successful: Iran and Saudi Arabia have made conflicting demands on Pakistani foreign policy, and even China has occasionally been irked by Pakistani support of Islamic militants in Xinjiang; the Pakistani-brokered peace accord between the *mujahideen* in Afghanistan has collapsed; and relations with the Central Asian states have turned out to be less useful than previously expected.

Given these developments, it is obvious that Pakistani options—never very bright to begin with—have been adversely affected in the aftermath of the Cold War. The loss of the United States as guarantor and ally cannot be compensated for very easily. And so Pakistani grand strategy will continue to salvage whatever it can of the relationship, making small compromises along the margin, but it is hard to imagine any abandonment of its nuclear program short of receiving iron-clad security guarantees. Pakistan will continue to offer its friendship to the United States (despite significant internal suspicions of U.S. regional policy), if for no other reason than to moderate any future U.S. support of India. But the spectacle of both India and Pakistan competing for American favor, diplomatic support, capital, technology, and weaponry will remain a conspicuous feature of South Asian geopolitics well into the foreseeable future.

LOOKING TO THE FUTURE: IMPLICATIONS FOR U.S. AIR POWER

For the first time since independence, India's democratic political structure appears ready and willing to sustain the economic liberalism required to expand material capabilities and, thereby, to enable it to seek true hegemonic status. This prognosis of success is justified mainly by the unique social coalitions supporting the present efforts at reform. Further, India already possesses the cultural framework to support expansion in the form of a liberal political, legal, and administrative framework—and the past 40-odd years have demonstrated both its durability and its flexibility. The prospects for success, therefore, appear propitious even if such success wears a distinctly Indian mask and is achieved amidst what looks like chaos from the outside. Consequently, the issue may no longer be *whether* India will be a hegemonic power, but rather *when* it will become one.

The situation in Pakistan is more complex; consequently, judgments are relatively uncertain. It is clear that significant segments of the elite population want to embark on real efforts at revitalizing Pakistani society in the face of the state. This, at any rate, is the real meaning of the present attempts at consolidating civilian political power and establishing a robust democratic polity. The issue, however, is more complex than simply confining the military to its barracks. It involves, among other things, resolving delicate problems of constitutional balance, developing a healthy political culture, and creating a consensus within the polity on the desired nature of the Pakistani state. It also involves proceeding purposefully with the program of economic liberalization—accompanied by reforms in the tax base, the modernization of agriculture, the liberation of critical industrial sectors from bureaucratic regulation, and the commitment to equity at the interprovince level. The efforts made thus far have been hesitant, and overall progress has been stymied by continual bickering among Pakistan's civilian elites in the face of continued resistance by the entrenched bureaucracies, a rapacious attitude on the part of the bourgeosie, and rising demands emanating from the increasingly vocal fringes of fundamentalist Islam. The prospects for eventual success are, therefore, uncertain because the objective of renewal is much more difficult to attain in Pakistan than it is in India. Pakistan is in many ways a beleaguered state; further, reform here is intended to achieve both economic revitalization *and* fundamental political restructuring. Consequently, the Pakistani program of renewal bears a much greater burden than the comparable effort occurring in India.

Very obviously, there are several factors that could retard the transformations currently occurring in both states, but perhaps the most important impediment is external: war. A conflict between India and Pakistan would upset the evolving reforms in more ways than can be imagined. At the very least, it would provide renewed justification for overt military intervention in politics in Pakistan; simultaneously, it would exacerbate fiscal pressures in India and, perhaps, derail the economic metamorphosis now under way for many years to come. If such a conflict involved the use of nuclear weapons, the resulting devastation could be substantial—depending on the kinds of weapons used and the targeting strategies employed. Moreover, any nuclear weapon use could accelerate a shift from what may be initially a conflict with limited aims to total war. The end result of such a conflict would be both difficult to control and impossible to foresee. For this reason, the current low-intensity conflict between India and Pakistan in Kashmir is extremely troublesome. It provides a focal point for an ongoing rivalry that could explode into open war under some as yet not clearly understood conditions.[17] However, it must be recognized that Kashmir is at best a manifestation and not a cause: Even if the Kashmir dispute were miraculously to disappear,

---

[17]RAND is currently conducting research aimed at understanding the conditions that make for deterrence stability in the Indian subcontinent. It is unfortunate that no serious study of this problem yet exists in the open literature. For a U.S. perspective on this problem, see Betts (1980).

Indo-Pakistani rivalry would not. Security competition between India and Pakistan arises because each state serves as an *objective* limitation on the goals, ambitions, and self-image of the other. This reciprocal dynamic produces a condition of mutual insecurity that episodically manifests itself in a variety of political contests but is not, ultimately, reducible to them.

Because Indo-Pakistani political competition is fundamental and *not* issue-driven, as is often imagined, it presents a very difficult challenge for American security and foreign policy. The kinds of U.S. responses that are desirable in this context cannot be analyzed here, but one crucial exception must be noted: It is important that the United States engage itself in South Asia simply to ensure that Indo-Pakistani security competition, even if conducted through war, never entails *any* nuclear weapon use. Their use in the Indo-Pakistani context would have three serious and, perhaps, unacceptable consequences for U.S. regional and strategic policy.

First, the use of nuclear weapons would provide increased incentives for expanded Chinese involvement in South Asia, possibly including a Chinese military intervention of some kind or another. Further, it would make U.S. allies in Southwest and Southeast Asia more insecure, simultaneously providing additional incentives to Iran and the Central Asian republics to nuclearize. Finally, and perhaps most importantly, it would break the taboo against nuclear use and encourage other states to acquire, deploy, and contemplate using the only class of weapons that could threaten U.S. security on a large scale. For this reason alone, the issue of possible nuclear use is of the utmost importance.

The nuclear capabilities of India and Pakistan do not directly threaten the continental United States today. Whether they threaten extended U.S. regional interests is debatable. In any case, U.S. air power serves to mitigate this threat—to the degree that it exists—simply by acting as a deterrent force as it did throughout the Cold War and continues to do. The really interesting question, therefore, is whether U.S. air power has a role in achieving the secondary goal of preventing India and Pakistan from attacking one another with nuclear weapons, given that the primary goal of preventing them from acquiring such weapons is now impossible to attain. It is possible to answer this question in the affirmative, however tentatively, and three classes of tasks can be suggested—very briefly—in this connection.

The first class of tasks involves developing and maintaining the requisite intelligence capabilities for understanding Indian and Pakistani conventional-nuclear capabilities and the ways in which the parties might intend and be able to use them. Understanding relative military capabilities, not just in terms of the raw numbers of weapon types but actually about the ability of these states to use their weapons effectively is fundamental, since it speaks to the issue of what kinds of military objectives could be secured (and, therefore, sought) by one or both sides in the context of a crisis under varying assumptions relating to strategic warning time, intensity of combat, and logistics and sustainability.[18]

The second class of tasks involves developing a range of contingency plans for using U.S. air power to deter or reassure one or both sides should policymakers choose to intervene militarily, either when conflict is imminent or already under way. This task involves planning the kind of force mix and capabilities required for a variety of contingencies ranging from providing airlift for a group of peacekeepers, through providing combat advice and intelligence support, to prosecuting the full range of warfighting operations associated with suppressing the mass destruction capabilities possessed by one or both states.

The third class of tasks involves planning for the disaster support and humanitarian relief that may be needed in the context of a South Asian war in which nuclear weapons are used. It is entirely possible that U.S. policymakers may reach the conclusion that American stakes in a South Asian conflict are too small to permit active military intervention for purposes of conflict termination.[19] Instead, U.S. intervention may be limited to providing disaster support in a permissive postconflict environment. These contributions may range from simple efforts, such as providing meteorological information for purposes of assessing fallout patterns, to more complex opera-

---

[18]For an excellent survey of the kind of intelligence and warning capabilities required, see Sokolski (1994).

[19]The nature of U.S. stakes in a South Asian nuclear conflict is explored in Millot et al. (1993), pp. 111–138.

tions, such as providing medical assistance, shelter, and communication support. Such operations could also be carried out in response to natural disasters, such as famine, flood, and earthquake, and could involve operations conducted throughout the South Asian region and not simply in India or Pakistan. Shortly after the Gulf War, the U.S. Army in fact was called upon to provide such support in Bangladesh (Operation Sea Angel) in the aftermath of a disastrous cyclone.[20] Should a disaster of similar or greater magnitude afflict the region, it is entirely possible that U.S. air power assets, particularly in the form of mobility and transportation capabilities, would be at the vanguard of U.S. outreach toward South Asia.

REFERENCES

Alam, Shahid, "Some Implications of the Aborted Sale of Russian Cryogenic Rocket Engines to India," *Comparative Strategy*, Vol. 13, No. 3, 1994, pp. 287–300.

Arquilla, John, "The Strategic Implications of Information Dominance," *Strategic Review*, Vol. 22, No. 3, 1994, pp. 24–30.

Bandhyopadyaya, J., *The Making of India's Foreign Policy*, Calcutta: Allied Publishers, 1980.

Basu, Tarun, "Russia Visit Termed a Success," *India Abroad*, July 8, 1994, p. 10.

Betts, Richard K., "India, Pakistan, and Iran," in Joseph Yager, ed., *Nonproliferation and U.S. Foreign Policy*, Washington, D.C.: Brookings, 1980, pp. 85–173, 323–365.

"China: The Titan Stirs," *The Economist*, November 28, 1992, pp. 3–5.

Cohen, Stephen P., "U.S.-Pakistan Security Relations," in Leo E. Rose and Noor A. Husain, eds., *U.S.-Pakistan Relations*, Berkeley: University of California Press, 1985, pp. 15–33.

Cohen, Stephen P., and Richard L. Park, *India: Emergent Power?* New York: Crane Russak, 1978.

---

[20] The nature of this operation is discussed at length in McCarthy (1994).

Fulghum, David A., "Indian Air Force Faces Tough Choices," *Aviation Week & Space Technology*, July 25, 1994, pp. 40–51.

Garrity, Patrick J., *Why the Gulf War Still Matters: Foreign Perspectives on the War and the Future of International Security*, Los Alamos: Center for National Security Studies, Report No. 16, 1993.

Kemp, Geoffrey, "A Roller Coaster Relationship: United States-Pakistan Relations After the Cold War," in David O. Smith, ed., *From Containment to Stability: Pakistan–United States Relations in the Post–Cold War Era*, Washington, D.C.: National Defense University, 1993, pp. 51–62.

McCarthy, Paul A., *Operation Sea Angel: A Case Study*, Santa Monica, Calif.: RAND, MR-374-A, 1994.

Mendis, Vernon L. B., *SAARC: Origins, Organisation, and Prospects*, Perth: Indian Ocean Center for Peace Studies, 1991.

Millot, Marc Dean, Roger Mollander, and Peter A Wilson, *"The Day After..." Study: Nuclear Proliferation in the Post–Cold War World:* Vol. II, *Main Report*, Santa Monica, Calif.: RAND, MR-253-AF, 1993.

Nayar, Baldev Raj, "A World Role: The Dialectics of Purpose and Power," in John W. Mellor, ed., *India: A Rising Middle Power*, Boulder, Colo.: Westview Press, 1979, pp. 117–179.

Nuri, Maqsudul Hasan, "Pakistan's Security Perceptions in the Post–Cold War Era," in Kanti P. Bajpai and Stephen P. Cohen, eds., *South Asia After the Cold War: International Perspectives*, Boulder, Colo.: Westview Press, 1993.

Panchamukhi, V. R., "Growth, Trade and Structural Changes: In the Asian Region," in Eric Gonsalves, ed., *Asian Relations*, New Delhi: Lancer, 1991, pp. 212ff.

Robinson, Francis, ed., *Cambridge Encyclopedia of India, Pakistan, Bangladesh, Sri Lanka, Nepal, Bhutan and the Maldives*, Cambridge: Cambridge University Press, 1989.

Sokolski, Henry, "Fighting Proliferation with Intelligence," *Orbis*, Vol. 38, No. 2, 1994, pp. 245–260.

Sundarji, Krishnaswamy, "Strategy: Objectives and Options," *India Today*, February 15, 1991.

Taqui, Jassim, "The Indo-American 'Defense Pact,'" *The Muslim*, Islamabad, October 21, 1991.

Tellis, Ashley J., "India's Naval Expansion: Reflections on History and Strategy," *Comparative Strategy*, Vol. 6, No. 2, 1987, pp. 185–219.

_____, "The Air Balance in the Indian Subcontinent: Trends, Constants and Contexts," *Defense Analysis*, Vol. 2, No. 4, 1986, pp. 264–268.

Tharoor, Shashi, *Reasons of State: Political Development and India's Foreign Policy under Indira Gandhi 1966–1977*, New Delhi: Vikas, 1982.

Chapter Thirteen

AFRICA

Margaret C. Harrell

INTRODUCTION AND BACKGROUND

The suffering and death of African peoples reached the forefront of American awareness in association with the U.S. deployment of troops to Somalia. This chapter offers a brief summary of the overall geopolitical status of sub-Saharan Africa and some snapshots of African states in crisis. A traditional assessment of U.S. strategic interests would probably exclude any possible U.S. military operations in sub-Saharan Africa in the immediate future. However, given the new era of international awareness and increasing peace operations, it is not clear that the United States will be able to remain apart from operations in Africa. Thus, this chapter assesses representative conflicts in Africa without saying where the United States might become involved and considers the demands that would be placed on U.S. forces, particularly the U.S. Air Force.

Political Summary

The number of conflicts in Africa is sobering. While the violence in some countries, such as Rwanda, Angola, Burundi, Somalia, and South Africa, is reported by Western media, other countries continue to have low-level conflicts, or potential conflicts, almost perpetually, unrecognized by much of the Western world. The Djibouti civil war continues; the Nigerian government was challenged by a coup; and the situation in the Ivory Coast is uncertain.[1] At the beginning of this decade, more than one hundred border disputes were ongoing in Africa.

Much of this conflict can be attributed to the lack of ethnic homogeneity across Africa and within individual countries. For example, Sudan includes people who speak a total of 115 languages and dialects (Somerville, 1990, p. 34). The differences in religion, which ranges from radical Islam to Christianity, to animalist beliefs, also separate the peoples, many of whom originally lived—or still live— tribal-nomadic life styles which do not contribute to defined international borders. This volatile mixture of unlike peoples is fueled by the apparent

> Failure of African regimes to legitimize or popularize their rule. . . . Far from seeking to alleviate the problems of ethnic multiplicity and conflict, regional rivalries, economies dependent on a few export crops and therefore on foreign markets, heads of state and whole governments tried to use the inheritance to entrench themselves in power and to enrich themselves. (Somerville, 1990, p. 186.)

Rather, African countries are largely ruled by force; at the beginning of the 1990s, half of the African countries were led by military governments.

Despite the well-known efforts of the United Nations (UN) to intervene, demobilize, and establish democratic governments in such countries as Angola, South Africa, and Mozambique, all indicators suggest that African conflicts will likely continue. The miserable living conditions of most Africans, the lack of political legitimacy discussed above, the perpetuity of a "winner takes all" mentality that encourages uprisings and coups (Somerville, 1990, p. 183), and the rapid population growth in Africa provide both the motivation and the manpower for continued conflict.[2] In addition, as much of this

---

[1] See Beaver (1994) for a brief discussion of the world's conflicts.

[2] Childress and McCarthy (1994, p. 8) have extracted trends from UN data that show that the population in less-developed areas will increase by as much as 75 percent between 1990 and 2025. This compares to a projected 12-percent increase in more developed areas for the same time period.

growth will occur in urban areas, youths become easy recruits for insurgent organizations with the resources to enable the conflicts.

Economic Summary

Economic conditions in Africa are extremely dismal and, even in the most wildly optimistic scenarios, will continue to remain so for quite a while. If the conflicts continue, the economic situation may not even meet the less-than-optimistic World Bank projection referred to in the following:

> The most dispiriting thing about Africa is not that it is the world's poorest continent; nor even that it is the only one where people were poorer at the end of the 1980s than they had been at the start. [E]ven if its economy (minus South Africa) were to grow at the rate confidently projected by the World Bank for the rest of the 1990s, Africans would have to wait another 40 years to clamber back to the incomes they had in the mid-1970s. Exclude Nigeria, and the wait would last a century.[3]

Not surprisingly, international interests have been very reluctant to invest in countries without stable and trustworthy governments and banking systems. And, as the economic situation continues to suffer, the public discontent will persevere.

SNAPSHOTS OF AFRICA

The large number of African countries and the variety of conflicts and problems throughout the continent complicate any attempt to summarize these issues. Several countries have been selected below for brief descriptions of their ongoing conflicts and potential problems at the time of this writing. These are intended to characterize and provide examples of the problems within African countries.

---

[3]"A Flicker of Light" (1994), pp. 21–24.

Somalia

The events of 1992–1993 familiarized Americans with Somalia, where starvation and anarchy were threatening the lives of multitudes. Although the situation in Somalia was not worse than that in other countries—such as Sudan—the United States responded with military force to provide security and assistance to the UN and humanitarian relief missions in Somalia. These efforts did considerable good in much of Somalia, but the overall mission was not successful. After the mission shifted to include disarmament and nation-building activities, which ran counter to the interests and ambitions of well-armed Somali factions, U.S. casualties and media coverage of the confusion between the United States and the UN soured the international effort in Somalia.

The developments in Somalia have grave implications for the likelihood of similar U.S. involvement in African nations and UN operations. The U.S. people now understand the risks involved in ill-defined or evolving operations in Africa. Other African leaders have also watched Somalia to ascertain the U.S. response to violence targeted against U.S. forces and now know that such violence will force the United States to reconsider its intervention in a conflict. In addition, poor projections of the future situation in Somalia dampen further any residual enthusiasm for similar operations.

When Siad Barre was overthrown in January 1991, the military systems and other capabilities acquired from such players as Libya provided the military means for conflict between numerous clan-based factions. Thousands were killed in 1991, and hundreds of thousands were displaced. There was no working government; most of the civilians were armed; and many were resorting to banditry to survive. Although reports following the U.S. deployment report sufficient quantities of food in Somalia, the distribution systems were not adequate to stop the starvation. In addition, country experts have asserted that the clans who did have food resources were not interested in or compelled to feed others, so access to food depended upon clan affiliations; the weaker did not eat.

The U.S. Operation Restore Hope (ORH) did secure the country and provide security for food and relief distribution. As ORH progressed, schools were rebuilt, people were vaccinated, judicial systems were reestablished, and governing councils were established. However, soon after the hand-off to UN command, the operation soured. While ORH had been a quick fix, none of the UN or U.S. operations in Somalia had taken the culture and motivations of Somalis into consideration. As the faction leaders began to feel threatened by the disarmament and emerging government policies that were being implemented, they withdrew their support for the UN operation. When the UN and the United States targeted a single faction leader, the other factions waited and observed while the UN became—to the people of Somalia—another faction in the fight, rather than a separate, impartial, adjudicator. In retrospect, an impartial solution to Somalia does not appear to have been feasible. Whomever the UN chose to help, they would disadvantage other rival clans. Thus, impartial assistance was not possible.

U.S. forces, including the diplomatic staff and the Marines left behind to provide them security, have all been withdrawn from Somalia. Many of the other countries have withdrawn their forces, and Somali violence against the remaining forces has continued to increase, prompting concern about the safety of a complete withdrawal from a country that views the UN equipment as "bounty" (Preston, 1994b, p. A29)—there for the taking if armed Somalis can overcome the UN forces in possession. All together, more than 100 peacekeepers were killed during the U.S. and UN efforts in Somalia, but the country is no closer to a political solution than it was before the deployment of the international force. Instead, because of both the salaries the UN and relief organizations paid to local Somalis and the large sums of money and other equipment stolen from the UN operation, some of the clans have been revitalized and are ready to continue their fighting.[4]

---

[4]The average Somali annual income was below US$100 before the UN operation. The *monthly* UN salary paid to Somalis ranged from several hundred to more than US$1,400. Data on the salaries paid by relief organizations are not available, but can be assumed to be even higher than those paid by the UN, as many of the guard positions were actually a form of extortion by the local clans.

Rwanda

The conflict in Rwanda has drawn worldwide attention since President Juvenal Habyarimana was killed when his aircraft (also carrying the Burundi president) went—or was shot—down on April 6, 1994. The Rwandan civil war had just ended with a signed peace deal in August of 1993 between the traditionally cattle-owning Tutsi minority and the traditionally peasant Hutu majority. Tutsi leadership had been established by the previous colonial government, but they were overthrown by the Hutus in 1959, and the bitterness between the two ethnic groups has continued since then.

The recent three-year civil war was well-armed by France, Egypt, and South Africa, but the Tutsi rebel Rwandan Patriotic Front used machetes and clubs as well as automatic rifles to capture two-thirds of the country and overthrow the government in the aftermath of April 6, 1994.[5] Hundreds of thousands (or possibly millions) of Hutus have fled to the far reaches of their country and into Tanzania and Zaire. The media-publicized horrors have included people hacked and clubbed to death and the survivors scrambling and begging for food in refugee camps completely inadequate for the masses of fleeing people. The deaths have taken a heavy toll on both ethnic groups, and the solution to this problem will not be a simple one. The status quo from April 6—the day of President Habyarimana's death—is too far removed from the current situation.

Although the Security Council "reluctantly" authorized 5,500 troops, most of the regular peace operation participants were reluctant to join France and the African countries in a mission that, with the vocal objections of the rebel forces, had no pretense of impartiality (Preston, 1994a). The United States initially responded to perceived international pressure to participate only by pledging airlift support and armored personnel carriers to the UN effort,[6] but then agreed to provide security and humanitarian support. This U.S. participation peaked at approximately 2,600 troops in August 1994, all of whom are scheduled to return home by the end of September 1994. While

---

[5]"Burundi and Rwanda: Joined in Death" (1994). Also see Goose and Smyth (1994, pp. 86–96) for an excellent account of the effect of small arms transfers to such countries as Rwanda.

[6]"U.S. Acting More Urgently to End Rwanda Slaughter" (1994, p. A12).

many are criticizing the late and limited response of the United States to the Rwandan crisis, the U.S. contribution to relief support in Rwanda is expected to reach a total of $500 million, approximately $2 for each U.S. citizen (Goose and Smyth, 1994, p. 88).

As of October 1994, the future of Rwanda is uncertain. More than one-half million people are estimated to have died—most of them the minority Tutsis. The number of refugees in nearby countries, such as Zaire and Tanzania, is enormous, and most of them are reluctant to return, given recent accounts of revenge-based torture perpetrated by the Tutsi. The economy of the country is dependent upon the return of the farming peasantry, most of whom are among the current refugees. Further, the departing Hutu government is reported to have taken boxes of Rwandan francs with them to Zaire. The resulting current scarcity of currency has crippled the efforts of the new government to return to normalcy. Any attempt to change the currency, however, would be stymied by the lack of operating banks and by the large number of absent people (Richburg, 1994, p. A46).

Zaire

Zaire is the second largest sub-Saharan African country. Belgium granted the Congo its independence in 1960, fully expecting that the Belgian administrators would continue to maintain control of the country. Instead, the Congolese Army mutinied against the Belgians, and requested the assistance and support of the UN against the reintroduction of Belgian troops. The mission was soon mired in a mess of internal Congo conflict, and the tremendous size of the country proved one of many difficulties the UN forces faced during that first UN "peace enforcement" operation in the 1960s.[7]

President Mobuto Sese Seko, Zaire's present ruler, began to establish his power in 1965. He dismissed the Parliament in 1966, established a new Constitution in 1967, and consolidated all of the state institutions into his political party, the Popular Movement of the Revolution, in 1971. Mobuto has remained firmly in power since

---

[7]Many peace-operation experts claim that the memory of the Congo mission kept the UN out of the African continent for such an extended period of time.

then, despite the several prime ministers he has appointed and dismissed if they proved more independent than he desired.

Despite significant natural resources, including copper and cobalt, Zaire has suffered continual economic problems, due in large part to extensive corruption and mismanagement by Mobuto, who has financed his personal fortune and extravagant lifestyle from the national coffers. Despite Mobuto's questionable leadership, Zaire enjoyed military aid and financial support during the 1980s from many Western countries, including France, Belgium, West Germany, the United States, Israel, and China.

Foreign support faded at the end of the 1980s, however, while internal unrest increased. In 1991, prodemocracy elements demanded Mobuto's resignation, and popular demonstrations ensued. Then, soldiers joined the rioters to protest the lack of pay; there was no money to pay their wages. An 80-percent devaluation of Zairian currency, an external debt of $10 billion, and an annual inflation rate of 1,500 percent had combined to cripple the economy. The continued riots prompted France and Belgium to organize evacuation operations to allow Westerners to leave Zaire.

Despite the introduction of a new currency in October 1993, the economic condition of the country has continued to falter, and Mobuto remains in power. The current conflict in Rwanda and the masses of refugees that have crossed Zaire's borders have continued to strain the resources of Zaire. One African expert maintains that the key difference between Zaire and other African nations, such as Somalia, is that Mobuto has kept such tight control over the country's borders and finances that the people did not have the resources to arm themselves. The future of Zaire, despite the relative lack of arms and munitions, is uncertain, and the vastness of the country would impede any external operation to assist the people or control the violence.

Angola

Conflict began in Angola even prior to its 1975 independence from Portuguese colonial rule. The government forces, the *Movimento Popular de Libertaçao de Angola* (MPLA), were supported by the Soviet Union and Cuba, while the United States and South Africa supported the rival *União Nacional para a Independência Total de Angola* (UNITA). During the 1980s, the situation worsened as South Africa conducted raids into Angola to destabilize the Angolan government. Cuba and South Africa withdrew in cooperation with the Namibian Accords at the end of the 1980s, but the 1989 peace initiatives failed and the conflict resumed. In 1990, Portugal mediated peace talks, supported by the United States and the Soviet Union, each eager to end their participation in a conflict that was of less interest in the absence of the Cold War. President Eduardo dos Santos (MPLA) and Jonas Savimbi (UNITA) signed a 1991 peace agreement that specified a cease fire, election plans, and a plan to create a single Angolan army from the two forces.

The elections conducted in September 1992 initially appeared successful—over 90 percent of the eligible voters participated, and the election, which was supervised by a small UN observer force, appeared a fair representation of the public.[8] However, when results indicated an MPLA victory, UNITA rescinded its support and declared the election fraudulent, despite the UN assertions to the contrary.

The fighting following the election—the demobilization had stalled and had not been completed prior to the election—was more deadly than that prior to the peace agreement. The United States recognized MPLA and lifted the ban of military sales to Angola in the summer of 1993, and the battles have since ravaged the country as each struggles to improve its bargaining position before it participates in additional peace talks. Both sides are spending lavish amounts on imported arms and ammunition; the government has oil revenues, and the rebels depend upon their income from smuggled diamonds. In addition, a government-employed force of more than 100 foreign mercenaries is battling.[9]

There is still a small UN observer force—less than 70 military and police observers—and some relief workers in Angola, but these personnel do not have access to the entire country and have been fired upon by both the government and the rebel forces. Both sides have

---

[8]See Fortna (forthcoming) for an excellent discussion of the events in Angola.

[9]"Angola: No Relief" (1994), p. 52.

also targeted the general population. The rebels shell the towns, and the government has dropped incendiary bombs into Huambo, the second largest Angolan city. The fighting will continue until the parties agree upon the role that UNITA will fulfill in any united government, and as each agreement nears, the fighting intensifies.

Nigeria

Since Nigeria became a federal republic in 1963, the government has vacillated between corrupt civilian leadership and corrupt military leadership. Coups and heavily rigged elections are the usual means of government transition. Unfortunately, the decades of corrupt leadership have left the country in massive debt. The budget deficit reached $4.7 billion, 15 percent of GDP last year, and the inflation is estimated at 100 percent a year.[10]

Since 1985, however, there have been governmental improvements in Nigeria. President Babangida worked to restore control over the country. His changes included moving the capital to reduce the Yoruba influence in the government; implementing infrastructure improvements, such as highways and telephones; creating additional states in the federation to increase representation and decrease ethnic tensions; and, perhaps most importantly, announcing tremendous economic reforms. Unfortunately, Babangida resigned in August 1993, just as the country was gaining a foothold on internal development and status among African nations.

The current political status of the country is uncertain. It is relatively calm now, but could erupt. The June 1993 elections were annulled by the government and the winner, Chief Moshood Abiola, is hiding from the threat of treason charges from the current military leader, General Sani Abacha. On the economic front, all improvements have ceased, and the World Bank has asserted that the new economic policies will certainly fail.[11]

Nigeria is especially interesting for its role within Africa; it has emerged as a recognized strength within Africa and a strong political

---

[10]"Nigeria Marches Backwards" (1994), p. 43.

[11]"Nigeria" 1994, p. 849.

element within the Organization of African Unity.[12] After President George Bush asked Nigeria to handle the Liberian situation (while the United States concentrated upon the Persian Gulf conflict), Nigeria succeeded in stabilizing Liberia.[13] The Liberia effort, however, has limited Nigeria's capability to respond to events within its own borders. Cameroon forces attacked Nigerian forces several times in February 1994, apparently trying to provoke a response. Nigeria has been reluctant, however, to mobilize forces for any response, which would likely involve withdrawing their stabilizing presence from Liberia.

South Africa

The world is watching South Africa, to see whether the recent elections and the end of *apartheid* will solve the country's problems. South Africa does not have all the same problems of other African countries; South Africa has low foreign debt, working legal and judicial systems, and an industrial infrastructure. The past pariah status of the country compelled it to develop its own defense capability, and South Africa is a self-declared prior nuclear power. However, the past has seen the benefits of these systems go largely to the ruling white minority, and it will become evident with time whether these benefits can be extended to the rest of the population.

South Africa has suffered for years from a tremendous amount of both political violence and basic crime. The bulk of the violence against blacks has been waged by other blacks, in ethnic and political extortion and battles of strength. As the 1994 elections neared, the political violence did not abate; rather, incidents among and involving members of the Inkatha party, which largely consists of the Zulu tribe, escalated as the Inkatha first declined to participate. Casualty figures ranged from 65 to 100 each week.

The constant and increasing violence inspired fear in the South African population. While most white citizens carry guns and live in houses with high electric fences and security systems, the security

---

[12]"Nigeria" 1994, p. 849.

[13]Unfortunately, Nigeria did so at the tremendous cost of at least US$500,000 a day, a huge strain upon its already struggling economy ("Nigeria," 1994, p. 848).

available in the township areas for black citizens is minimal. The influence of the peace organizations and the observers, however, is notable. Despite their small numbers, the presence of representative observers from the UN, the EC, and other domestic and international organizations has calmed many of the public gatherings which might otherwise have escalated into violence.

The status of South Africa is unclear. Nelson Mandela, new South African President and prior head of the African National Congress, has promised to improve the quality of life of black South Africans and to return land to its previous black owners. The economic means to complete these tasks quickly, however, is not immediately clear. Whether the people will wait patiently as equality progresses slowly is also unclear. In fact, many of the advisors with whom Mandela has surrounded himself know little about how to run a government.

The future of South Africa depends upon the expertise and finesse of the new government as it struggles to fulfill a myriad of expectations among the black South Africans, the willingness of the white South Africans to participate fully in the transitional phases of government, and the attitudes of various black ethnic groups, most notably the Inkatha, as they are governed by other blacks. Whether the outcome is stability or violence, South Africa will have a strong influence on the rest of Africa, especially Southern Africa, where either extreme—wealthy South African tourists or outreaching violence—could affect the future of countries like Mozambique.

Mozambique

The international community is watching Mozambique for signs of whether the UN-sponsored elections will actually occur, much less whether they will transform Mozambique from a war-riddled shell into a sustainable democracy. Independence was not granted to Mozambique until 1975, and the local people were ill-equipped to handle the independence once it finally occurred. As the Portuguese colonists left, the inadequacies and inabilities of the local people to administer the country became woefully evident, as the Portuguese had done little to develop an indigenous elite, and the literacy rate was only 10 percent.

The Front for the Liberation of Mozambique (FRELIMO) was handed control by the Portuguese, and it soon became active in external African events by supporting guerrillas operating against neighboring Rhodesia (now Zimbabwe). When the Rhodesians retaliated with ground and air strikes in 1978 and 1979, Soviet and Cuban assistance provided Mozambique with a military force, including tanks, helicopters, and SA-7 surface-to-air missiles. Cuban, Soviet, and East German advisors were also resident in Mozambique, providing training assistance.

Meanwhile, the Mozambique National Resistance (RENAMO) was established by Rhodesians to monitor events within Mozambique.[14] RENAMO grew into a large insurgency movement, and began taking control of rural Mozambique. The war between RENAMO and FRELIMO progressed for sixteen years, displacing at least a million people, and killing hundreds of civilians at a time. The conflict compounded other almost insurmountable problems, such as repeated droughts, inadequate administration, and the crumbling infrastructure.

The late 1980s were characterized by discussions between the two parties, punctuated by RENAMO violence. Eventually, they agreed in December 1990 to end the war and merge their military forces. However, a cease fire was not successfully implemented until October 1992. The two parties agreed to demobilize under UN guidance and create a restructured smaller force comprised of equal numbers from each side. An election was planned and scheduled upon completion of the demobilization.

Currently, there are approximately 6,000 UN peacekeeping troops and several hundred unarmed observers in Mozambique. The peacekeeping troops have successfully secured the roads that run through Mozambique and connect its neighboring countries with the coast. However, the demobilization and election process is far behind schedule and is characterized by stalling and delays on both sides. As of this writing, the election is optimistically planned for October 1994. Meanwhile, demining has begun to clear millions of mines from the countryside of Mozambique, and there are millions

---

[14]"Mozambique" 1994, p. 787.

of people to resettle. Should the losing party accept the loss and begin only to prepare for the next election, Mozambique could be a UN success story. Regardless, a successful demobilization is crucial to any stable future in Mozambique.

U.S. INTERESTS IN AFRICA

Unfortunately, there currently is no clear policy for U.S. military intervention in Africa. Although *The Clinton Administration's Policy on Reforming Multilateral Peace Operations* (U.S. Department of State, 1994) is not Africa-specific, it is possibly the best current policy statement regarding potential U.S. intervention. It notes that

> territorial disputes, armed ethnic conflicts, civil wars (many of which could spill across international borders) and the collapse of governmental authority in some states are among the current threats to peace . . .

and notes that their cumulative effect may be significant. However, it still refers any intervention decisions back to the as-yet undefined American interests in the region.

The following paragraphs discuss various potential U.S. interests in Africa that might provoke military intervention or U.S. support of military operations in Africa. Any of these interests, of course, are balanced not only by competing domestic issues, but also by the recent memory of American lives lost in Somalia for what seemed at the time a simple "humanitarian operation."

Safety and Security of U.S. Citizens

This includes two separate concerns. First, the safety of any U.S. citizens in Africa is always of concern. Although the United States does not have the large population centers in Africa like those of former colonial powers, there are embassies and consulates throughout the continent, many of which could require assistance should they find themselves in the midst of a worsening conflict. Second, the United States will continue to act to prevent terrorist acts against U.S. citizens anywhere in the world.

Growth of Democratic Governments with Self-Sustaining Economies

Optimists would associate the prevention of many of the conflicts, which result in starvation and brutal deaths, with the growth of democratic governments with self-sustaining economies. U.S. military operations, such as special training programs, have worked for years to promote democratic values and provide valuable skills to and through the militaries of many African countries.

Conflict Solution and Prevention

While it will likely continue to pull at American heartstrings when the media reports starving children and mass murders, it is not clear which or how many of these cases fall directly within U.S. interests. In addition, competing domestic priorities, lack of sustainable U.S. support, and the potential cost in American lives lost will continue to limit the inclusion of these issues in any definition of U.S. interests.

Viability and Credibility of the UN and Other Regional Organizations in Africa

The United States cannot afford for the UN to lose credibility. This issue may provide the impetus for U.S. support of, and even involvement in, various operations within Africa.

Improved Capability of African Militaries Who Participate in Peace Operations

When UN- or otherwise-sponsored military operations employ the troops of African nations, train them, and provide them with practical experience, these troops become increasingly prepared for additional operations. Thus, when the United States supports relief or peace operations, even in locations of only marginal interest to the United States, we enable African nations to train and prepare for additional operations that may be more important to U.S. interests. Liberia is an excellent example of a case in which African countries successfully took the lead in peace operations. There is, however, the serious risk that these troops will put their newly acquired or refined skills and materiel to ill use in their own country.

Strategic Materials

The United States is very dependent upon several materials mined in Africa, including chromium, cobalt, manganese, platinum group metals, and vanadium. Of these, Africa is basically the sole source of chromium and cobalt, and Zaire alone produces 58 percent of the cobalt (see Allen and Noehrenberg, 1992). While these materials are of strategic importance to the United States, it is difficult to foresee a situation in which all the producing mines of either or both substances are closed, and the U.S. strategic stockpiles are simultaneously depleted. While these materials are of great importance to U.S. defense and manufacturing industries, a shortage of these materials, without any extraordinary circumstances, is unlikely to provoke a U.S. military operation.

The U.S. response in defense of these or other interests in Africa is, at best, unpredictable. International news coverage, special interest groups, and simultaneous events will influence the decision about whether or not to intervene in any instance. For example, although the United States intervened to end the starvation and suffering in Somalia, there were more displaced persons (and thus more of the accompanying problems, such as starvation) in Mozambique and South Africa; in Sudan, there was almost three times the displaced population of Somalia.[15] If the proposed U.S. economic and military assistance to a single nation is an indicator of the degree of U.S. interest in that country, then the U.S. interest in Somalia, expressed in proposed FY93 assistance determined prior to the U.S. military involvement in Somalia, was much less than that of at least seven other African nations (Childress and McCarthy, 1994, p. 31). Thus, the U.S. intervention in Somalia must have been influenced by unpredictable factors, which certainly included media pressure and possibly included avoidance of Yugoslavia, which was perceived at the time to be "messier." While there will likely be greater hesitancy

---

[15]See Childress and McCarthy (1994) for an excellent extrapolation of the available data sources and insightful discussion of the military implications of these demographic trends.

and reluctance to become involved in African crises in the "post-Somalia" future, the predictability of U.S. intervention, or support of intervention operations, may remain low.

POTENTIAL TYPES OF U.S. OPERATIONS IN AFRICA

The range of potential military operations for which the United States might provide support or employ forces is discussed in the following paragraphs.

Evacuation Operations

As long as the United States maintains embassies and consulates in countries characterized by uncertain stability, the likelihood remains that the United States may have to conduct military operations to evacuate American citizens. These operations could resemble the evacuation operation conducted from Liberia, in which American diplomats and their families were rescued. Evacuation operations might, however, resemble the U.S. operation on Grenada, in which a larger number of civilians, without any direct contact or communications with the U.S. military, needed to be evacuated. In addition, the United States might be requested to assist in the evacuation of European citizens, who reside in Africa in greater numbers than U.S. citizens. These operations are difficult to predict, but are usually conducted very quickly and require little sustainment.

Humanitarian Relief

These operations include only those relief operations without a security component and would most likely be caused by a natural disaster. For example, while Operation Sea Angel, the relief operations conducted by the U.S. Marines in Bangladesh, would qualify as a humanitarian relief operation, the operations conducted in Somalia would not. The U.S. involvement could range from airlift of forces or air delivery of supplies to actual troops on the ground distributing relief. Given the currently tainted public perception of "humanitarian operations" resulting from the use of that description as an early misnomer of the Somalia operations, it is likely that the United States would limit its involvement in such an operation to air support with minimal ground troop involvement. It is worthwhile to distinguish between relief operations resulting from a single natural disaster, such as a hurricane or an earthquake, and those resulting from an ongoing ecological disaster, such as drought, erosion, or the encroachment of desert sands. While the first is conceivably a temporary and "fixable" situation, the second can be relieved, but not "fixed," and will not be as simple to terminate.

Peace Operations

This category represents an assortment of operations, including observer missions (both traditional observer missions, such as the UN Troop Supervisory Operation mission, and newer election-monitoring and demobilization missions, such as the South Africa and Mozambique missions); traditional peacekeeping, such as the UN operation in Cyprus; and peace enforcement, such as the Somalia operations. Any U.S. role in the smaller observer missions would likely involve only a small number of personnel, such as those currently stationed in the Western Sahara. The traditional peacekeeping missions are not likely to require any military capability unique to the United States—peacekeeping operations have been ongoing for decades without the direct involvement of U.S. forces—but may require troops for political reasons. The peace enforcement missions are the most likely to require the expertise and materiel of U.S. forces. These missions, however, are also the most likely to resemble the ill-fated Somalia operation with a potentially volatile environment. Thus, peace-enforcement missions are the least likely to receive domestic and congressional support. Considerable international pressure or a threat to the viability of a UN operation could, however, compel the United States to become involved in any of these types of operations.[16]

This being the case, U.S. airlift for UN or multinational troops and logistics is a likely contribution that would raise the least objections in Congress and from the public. Should the concern about and wariness of such operations decrease, the U.S. psychological opera-

---

[16]It is noteworthy that, despite the lack of U.S. interests in Rwanda, and despite the memories of the Somalia operations, the United States has contributed airlift to the UN mission in Rwanda.

ions (PSYOPs), civil affairs, and military police personnel offer the capabilities most needed in these operations and lacking in other countries military forces, but due to the limited active-duty resources in these areas, infantry personnel would likely deploy.

Traditional Special Operations

U.S. special forces will likely continue to conduct such operations as internal defense and development (IDAD) training through the International Military Education and Training program throughout Africa.[17]

CONCLUSIONS

The geopolitical situation of Africa is extremely unstable, with dismal economic forecasts, and continuous conflict in many countries. The United States does not have many overwhelming interests in Africa— with the possible exception of the safety and security of U.S. citizens—that would result in any large scale, long-term unilateral operation.

U.S. operations in Africa are likely to fall within one of three categories. First, the United States will probably be involved to some degree with humanitarian relief and peace operations in Africa. Given the current negativity regarding any operation vaguely similar to that conducted in Somalia, the United States will probably limit its participation in any peace operations to some degree of support for other countries contributing contingents to the force. These operations will usually be conducted multilaterally, and they will be long and difficult to terminate. Second, such efforts as special operations in Africa are likely to continue. While the value of some of these operations is somewhat uncertain, they receive little public scrutiny and do not compromise the remaining deployable U.S. military capability.[18] The third group of operations includes evacuation, retaliation, and strikes against weapons of mass destruction. These

---

[17]IMET is a subcomponent of the Security Assistance and Training Program and "is designed to enhance the proficiency, professional performance, and readiness of foreign armed forces," U.S. Army (1993).

[18]See McCoy (1994) for a comparison of the IDAD operations in Senegal and Liberia.

operations are difficult to predict, require quick response once the have been decided, and are potentially crucial to the safety and well being of American citizens. In addition, these operations may depend heavily upon high technology support.

REFERENCES

"A Flicker of Light," *The Economist*, March 5, 1994, pp. 21–24.

Allen, Patrick D., and Peter C. Noehrenberg, *U.S. Dependence on Strategic Materials from Southern Africa Nations*, Santa Monica, Calif.: RAND, R-4165-OSD, 1992.

"Angola: No Relief," *The Economist*, June 18, 1994, pp. 51–52.

Beaver, Paul, "Flash Points Review," *Jane's Defense Weekly*, January 8, 1994, pp. 15–21.

"Burundi and Rwanda: Joined in Death," *The Economist*, April 9, 1994.

Childress, Michael T., and Paul A. McCarthy, *The Implications for the U.S. Army of Demographic Patterns in the Less Developed World*, Santa Monica, Calif.: RAND, MR-256-A, 1994.

Fortna, Virginia Page, "Success and Failure in Southern Africa: Peacekeeping in Namibia and Angola," in Donald C. F. Daniel and Bradd C. Hayes, eds., *Beyond Traditional Peacekeeping*, London: The Macmillan Press, forthcoming.

Goose, Stephen D., and Frank Smyth, "Arming Genocide in Rwanda," *Foreign Affairs*, September–October 1994, pp. 86–96.

McCoy, Jr., William H., *Senegal and Liberia: Case Studies in U.S. IMET Training and Its Role in Internal Defense and Development*, Santa Monica, Calif.: RAND, N-3637-USDP, 1994.

"Mozambique," *Defense & Foreign Affairs Handbook*, London: International Media Corporation Ltd., 1994.

"Nigeria," *Defense and Foreign Affairs Handbook*, London: International Media Corporation Ltd., 1994.

Nigeria Marches Backwards," *The Economist*, February 26, 1994, p. 43.

Preston, Julia, "U.N. Backs French Move Into Rwanda," *Washington Post*, June 23, 1994a, p. 1.

_____, "U.S. Troops May Aid in U.N. Withdrawal from Somalia," *Washington Post*, September 16, 1994b, p. A29.

Richburg, Keith B., "Rwandan Leaders Struggle to Rebuild Nation," *Washington Post*, September 25, 1994, p. A46.

Somerville, Keith, *Foreign Military Intervention in Africa*, London: Pinter Publishers, 1990.

"U.S. Acting More Urgently to End Rwanda Slaughter," *New York Times*, June 16, 1994, p. A12.

U.S. Army, *Operations*, Army Field Manual 100-5, June 14, 1993.

U.S. Department of State, *The Clinton Administration's Policy on Reforming Multilateral Peace Operations*, May 1994.